Crafts studies
Series editor: N. Hiller

ELECTRICAL CRAFT PRINCIPLES
Volume 1

ELECTRICAL CRAFT PRINCIPLES

Volume 1 – 3rd Edition

J.F. Whitfield, M.C.I.B.S.E., C. Eng., M.I.E.E.
Senior Lecturer in Electrical Engineering
Norwich City College of Further & Higher Education
Norwich
England

Peter Peregrinus Ltd. on behalf of the Institution of Electrical Engineers

Published by Peter Peregrinus Ltd., London, United Kingdom

First published 1974
© 1974: Peter Peregrinus Ltd.
Revised 2nd edition
© 1980: Peter Peregrinus Ltd.
Reprinted with minor corrections 1986
Reprinted 1987 (twice)
© 1988: 3rd edition Peter Peregrinus Ltd.
Reprinted 1990, 1993

ISBN: 0 86341 153 3

Apart from any fair dealing for the purposes of research or private study, or criticism or review, as permitted under the Copyright, Designs and Patents Act, 1988, this publication may be reproduced, stored or transmitted, in any forms or by any means, only with the prior permission in writing of the publishers, or in the case of reprographic reproduction in accordance with the terms of licences issued by the Copyright Licensing Agency. Inquiries concerning reproduction outside those terms should be sent to the publishers at the undermentioned address:

Peter Peregrinus Ltd.,
Michael Faraday House,
Six Hills Way, Stevenage,
Herts. SG1 2AY, United Kingdom

While the author and the publishers believe that the information and guidance given in this work is correct, all parties must rely upon their own skill and judgment when making use of it. Neither the author nor the publishers assume any liability to anyone for any loss or damage caused by any error or omission in the work, whether such error or omission is the result of negligence or any other cause. Any and all such liability is disclaimed.

Printed in England by Short Run Press Ltd., Exeter

Preface

This is the first of two volumes which together set out to cover the Electrical Principles syllabuses for the City & Guilds 200/231 (500 Series) Craft Studies Courses in engineering. The present volume covers all the work required for Part I of this course, and will also meet the requirements of Part I of the Electrical Craft Practice Course 181(347), the Electricians A Certificate 185(51A), and the first year of the Electricians B Certificate Course 231(51B). It should also prove useful as a first course in self study for practising and would-be tradesmen in electrical and allied crafts.

In general, the units used conform to the requirements of the SI and of the Council of Technical Examining Bodies. Notable exceptions are the units for magnetomotive force and magnetising force, where the 'ampere-turn' and the 'ampere-turn per metre' have been retained. In the opinion of the author, these terms have more meaning to the trade student than the 'ampere' and the 'ampere per metre'. It is felt that to continue to use Imperial units simply because we are familiar with them will prolong the agony of changing to the new and superior unit system.

All line diagrams have been deliberately reduced to their simplest forms, so that a student could reproduce them for examination purposes if necessary. In general, symbols conform to BS3939.

Solutions are to three significant figures, except where the work calls for greater accuracy. At the end of each chapter is a collection of the important formulas occurring during the chapter. Mostly these are expressed several times with different quantities as subjects.

Every care has been taken to ensure accuracy, but the author will be grateful for the notification of errors.

Norwich   J.F. Whitfield

Preface to Second Edition

This second edition has been made necessary by some changes in the structure of technical education. It conforms to the City & Guilds of London Institute 236 syllabus for electricians, Volume I covering Part I of that syllabus.

The first edition has been used for Part I City & Guilds Technicians, particularly the Electrical (280) and Electrical Installation (280) series. These courses have now been replaced by the TEC (Technician Education Council) system, and this volume will form a good basis for many Electrical Principles units at levels I and II, as well as for Electrical Applications at various levels.

Norwich   J.F. Whitfield

Preface to Third Edition

Important and far-reaching changes in the terminology applying to electrical installations have come about as a result of the introduction of the 15th edition of the I.E.E. Wiring Regulations. Whilst these will not affect all craft students, a significant proportion will need to use the current terminology as part of their studies and of their work. The book has been updated to take account of these changes.

Modifications have also been made to the standard First Certificate and early N level units of the Ordinary National Certificate of the Business and Technician Education Council (BTEC). Account has been taken of these changes to allow the book to continue to be suitable for use in such courses.

Norwich July, 1988   J.F. Whitfield

Acknowledgments

None of the material in this book can be claimed as original, and the author acknowledges with thanks the many teachers, students and writers who have collectively been responsible for this approach to the subject. In particular, he extends grateful thanks to

- The Institution of Electrical Engineers for permission to quote from their Wiring Regulations.
- The City & Guilds of London Institute (C & G), the Northern Counties Technical Examinations Council (NCTEC) and the Union of Lancashire & Cheshire Institutes (ULCI) for permission to publish questions from past examination papers. Where solutions are given to such questions, these are the sole responsibility of the author.
- P.M. Dunne, C. Eng., A.F.R.Ae.S., for assistance with some of the drawings
- M. Hannaford for assistance with some of the photographs
- N. Hiller, B.Sc. (Eng), C.Eng., F.I.E.E., the Editor, for his encouragement and courteous assistance
- Mrs. Joan Crockford for interpreting and typing the manuscript
- his wife, for her patient help and understanding
- his daughter Lesley, for typing the new material for the second edition
- the following organisations, which gave valuable assistance with information and illustrations:

    Alkaline Batteries Ltd.
    BICC Ltd.
    BICC-Burndy Ltd.
    Dorman & Smith Switchgear Ltd.
    English Electric Co.
    Erie Electronics Ltd.
    General Electric Co.
    Mallory Batteries Ltd.
    M. K. Electric Ltd.
    Ottermill Switchgear Ltd.
    Parkinson Cowan Heating Ltd.
    Power Centre Co.
    British Telecom plc
    Satchwell Controls Ltd.

Contents

Contents of chapters		page viii
Tables		xiii
Symbols and abbreviations		xiv
SI units		xvii
Basic electric units and circuits	chapter 1	1
Resistance and resistors	2	23
Mechanics	3	41
Heat	4	57
Electrical power and energy	5	65
Permanent magnetism and electromagnetism	6	77
Applications of magnetism	7	89
Electric cells and batteries	8	99
Electromagnetic induction	9	115
Basic alternating current theory	10	125
Electric motor principle	11	145
Practical supplies and protection	12	153
Cables and enclosures	13	173
Lighting and heating installations	14	189
Introduction to electronics	15	203
Numerical answers to exercises		212
Index		216

Contents of chapters

1 Basic electric units and circuits

page

1.1	Simple electron theory	2
1.2	Quantity of electricity and unit of current	3
1.3	Effects of electric current	4
1.4	Electric conductors and insulators	5
1.5	Electrical energy, work and power	7
1.6	Electromotive force and potential difference	8
1.7	Resistance: Ohm's law	9
1.8	Electric circuit	10
1.9	Ammeters and voltmeters	10
1.10	Series circuits	11
1.11	Parallel circuits	13
1.12	Series-parallel circuits	16
1.13	Summary of formulas for Chapter 1	19
1.14	Exercises	20

2 Resistance and resistors

2.1	Introduction	24
2.2	Effect of dimensions on resistance	25
2.3	Resistivity	29
2.4	Resistance calculations	30
2.5	Effect of temperature on resistance	32
2.6	Effects of temperature changes	34
2.7	Voltage drop in cables	35
2.8	Summary of formulas for Chapter 2	37
2.9	Exercises	37

		page
3	**Mechanics**	
3.1	Mass, force, pressure and torque	42
3.2	Work, energy and power	44
3.3	Lifting machines	46
3.4	Power transmission	49
3.5	Parallelogram and triangle of forces	51
3.6	Summary of formulas for Chapter 3	54
3.7	Exercises	54
4	**Heat**	
4.1	Heat	58
4.2	Temperature	58
4.3	Heat units	59
4.4	Heating time and power	60
4.5	Heat transmission	61
4.6	Change of dimensions with temperature	62
4.7	Summary of formulas for Chapter 4	63
4.8	Exercises	64
5	**Electrical power and energy**	
5.1	Units of electrical power and energy	66
5.2	Electromechanical conversions	71
5.3	Electric heating	72
5.4	Summary of formulas for Chapter 5	74
5.5	Exercises	74
6	**Permanent magnetism and electromagnetism**	
6.1	Magnetic fields	78
6.2	Units of magnetic flux	79
6.3	Electromagnet	80
6.4	Calculations for aircored solenoids	82
6.5	Effect of iron on magnetic circuit	84
6.6	Permanent magnets	86

			page
	6.7	Summary of formulas for Chapter 6	86
	6.8	Exercises	86

7 Applications of magnetism

	7.1	Introduction	90
	7.2	Bells and buzzers	90
	7.3	Bell indicators and circuits	91
	7.4	Relays and contactors	92
	7.5	Telephones	94
	7.6	Simple telephone circuits	95
	7.7	Loudspeakers	96
	7.8	Moving-iron instruments	97
	7.9	Exercises	98

8 Electric cells and batteries

	8.1	Storing electricity	100
	8.2	Primary cells	100
	8.3	Secondary cells	102
	8.4	Care of secondary cells	104
	8.5	Internal resistance	106
	8.6	Batteries	109
	8.7	Capacity and efficiency	111
	8.8	Summary of formulas for Chapter 8	112
	8.9	Exercises	112

9 Electromagnetic induction

	9.1	Introduction	116
	9.2	Dynamic induction	116
	9.3	Relative directions of e.m.f., movement and magnetic flux	118
	9.4	Simple rotating generator	119
	9.5	Direct-current generator	120
	9.6	Static induction	122
	9.7	Summary of formulas for Chapter 9	122

		page
9.8	Exercises	123

10 Introduction to alternating current

10.1	What is alternating current?	126
10.2	Advantages of a.c. systems	127
10.3	Values for a.c. supplies	127
10.4	Sinusoidal waveforms	129
10.5	Phasor representation and phase difference	131
10.6	Resistive a.c. circuit	133
10.7	Inductive a.c. circuit	134
10.8	Capacitive a.c. circuit	135
10.9	Transformer	137
10.10	Summary of formulas for Chapter 10	140
10.11	Exercises	141

11 Electric motor principle

11.1	Introduction	146
11.2	Force on a current-carrying conductor lying in a magnetic field	146
11.3	Relative directions of current, force and magnetic flux	147
11.4	Lenz's law	148
11.5	Direct-current-motor principles	149
11.6	Moving-coil instrument	150
11.7	Summary of formulas for Chapter 11	151
11.8	Exercises	151

12 Practical supplies and protection

12.1	Introduction	154
12.2	Direct-current supplies	154
12.3	Single-phase a.c. supplies	154
12.4	3-phase a.c. supplies	154
12.5	Earthing	156
12.6	Fuses	159
12.7	Circuit breakers	161

			page
12.8	Risk of fire and shock		164
12.9	Polarity		165
12.10	Safety precautions		166
12.11	Regulations		167
12.12	Electric shock		168
12.13	Artificial respiration		169
12.14	Summary of formulas for Chapter 12		169
12.15	Exercises		170

13 Cables and enclosures

13.1	Introduction	174
13.2	Conductor materials and construction	174
13.3	Cable insulators	175
13.4	Bare conductors	176
13.5	Plastic- and rubber-insulated cables	177
13.6	Sheathed wiring cables	177
13.7	Mineral-insulated cables	178
13.8	Armoured cables	179
13.9	Cable joints and terminations	180
13.10	Conduits	183
13.11	Ducts and trunking	185
13.12	Cable ratings	185
13.13	Exercises	187

14 Lighting and heating installations

14.1	Introduction	190
14.2	Mains equipment	190
14.3	Lighting circuits	192
14.4	Socket-outlet circuits	194
14.5	Other circuits	196
14.6	Earthing and polarity	197
14.7	Simple testing	198

			page
	14.8	Exercises	200
15		**Introduction to electronics**	
	15.1	Introduction	204
	15.2	Resistors for electronic circuits	204
	15.3	Semiconductor diode	208
	15.4	Semiconductor-diode types	209
	15.5	Exercises	209

Tables

1	Some devices relying on effects of electric current	5
2	Electrical conductors	6
3	Electrical insulators	6
4	Resistivity of common conductors	30
5	Temperature coefficients of resistance of some conductors	33
6	Specific heats	59
7	Standard sizes of copper cables with comparable aluminium sizes	**174**
8	Stranding of copper cables	**175**

Symbols and abbreviations

1 Terms

Term	Symbol or abbreviation
alternating current	a.c.
area	A or a
area, cross-sectional	c.s.a.
capacitance	C
charge (electrical)	Q
current	
steady or r.m.s. value	I
instantaneous value	i
maximum value	I_m
average value	I_{av}
distance	d
electromotive force	e.m.f.
steady or r.m.s. value	E
instantaneous value	e
energy	W
force	F
frequency	f
inductance, self	L
length	l
line current	I_L
line voltage	V_L
magnetic flux	Φ (phi)
magnetic flux density	B
magnetising force	H
magnetomotive force	m.m.f.
mass	m

Term	Symbol or abbreviation
mechanical advantage	m.a.
permeability of free space	μ_0 (mu)
permeability, relative	μ_r (mu)
phase angle	ϕ (phi)
phase current	I_P
phase voltage	V_P
potential difference	p.d.
steady or r.m.s. value	V
instantaneous value	v
maximum value	V_m
average value	V_{av}
power	P
primary current	I_1
primary e.m.f.	E_1
primary turns	N_1
primary voltage	V_1
quantity of electricity	Q
reactance, capacitive	X_c
reactance, inductive	X_L
resistance	R
resistivity	ρ (rho)
revolutions per minute	rev/min
revolutions per second	rev/s
root mean square	r.m.s.
rotational velocity	ω (omega)

Term	Symbol or abbreviation	Term	Symbol or abbreviation
secondary current	I_2	temperature coefficient of resistance	α (alpha)
secondary e.m.f.	E_2	time	t
secondary turns	N_2	time, periodic	T
secondary voltage	V_2	torque	T
speed (rev/min)	N	velocity ratio	v.r.
speed (rev/s)	n		

2 Units

Unit	Symbol	Unit of
ampere	A	electric current
ampere-turn	At	magnetomotive force
coulomb	C	electrical charge or quantity
cycle per second (or hertz)	c/s (or Hz)	frequency
degree Celsius	°C	temperature
henry	H	inductance
hertz	Hz	frequency
joule	J	work and energy
kilogram	kg	mass
kilowatt	kW	power
kilowatt-hour	kWh	energy
metre	m	length
millimetre	mm	length
newton	N	force
newton metre	Nm	torque
metre newton (joule)	mN	work and energy
ohm	Ω (omega)	resistance
radians per second	rad/s	rotational velocity
second	s	time
tesla	T	magnetic flux density
volt	V	p.d. and e.m.f.

Unit	Symbol	Unit of
watt	W	power
weber	Wb	magnetic flux
weber per square metre (or tesla)	Wb/m² (or T)	magnetic flux density

3 Multiples and sub-multiples

Prefix	Symbol	Meaning	
meg- or mega-	M	one million times	or $\times 10^6$
kil- or kilo-	k*	one thousand times	or $\times 10^3$
milli-	m	one-thousandth of	or $\times 10^{-3}$
micro-	μ (mu)	one-millionth of	or $\times 10^{-6}$
nano-	n	one-thousand-millionth of	or $\times 10^{-9}$
pico-	p	one-million-millionth of	or $\times 10^{-12}$

* may be changed in due course to K

SI Units

If we are to measure physical, mechanical and electrical quantities, we must use a system of units for the purpose. All units, no matter how complex, are based on a number of basic units. The system of units adopted in much of the world is the SI (an abbreviation of Système International d'Unités). Basic SI units:

Physical quantity	Name of unit	Symbol
length	metre	m
mass	kilogram	kg
time	second	s
electric current	ampere	A
temperature	kelvin	K
luminous intensity	candela	cd

Owing to the wide adoption of SI units, and the need for its complete use as early as possible, no mention has been made of the older units and unit systems in this volume.

Multiple and submultiple units

There are many examples in practical electrical engineering where the basic units are of inconvenient size.

Multiple units are larger than the basic units.

The prefix **meg** or **mega** (symbol M) means **one million times**.

For instance, 1 megavolt (1 MV) = 1 000 000 volts

and 1 megohm (1 MΩ) = 1 000 000 ohms

The prefix **kil** or **kilo** (symbol k) means **one thousand times**.

For instance, 1 kilovolt (1 kV) = 1 000 volts.

It seems likely that, in due course, the symbol K will be used in place of k.

Submultiple units are decimal fractions of the basic units.

The prefix **milli** (symbol m) means **one-thousandth of**.

For instance, $1 \text{ milliampere (1 mA)} = \frac{1}{1\,000}$ ampere

The prefix **micro** (symbol μ, the Greek letter 'mu') means **one-millionth of**.

For instance, $1 \text{ microhm (1 } \mu\Omega) = \frac{1}{1\,000\,000}$ ohm

The prefix **nano** (symbol n) means **one-thousand-millionth of**.

For instance, $1 \text{ nanosecond (1 ns)} = \frac{1}{1\,000\,000\,000}$ second

The prefix **pico** (symbol p) means **one-million-millionth of**.

For instance, $$1 \text{ picovolt } (1\,\text{pV}) = \frac{1}{1\,000\,000\,000\,000} \text{ volt}$$

Two words of warning are necessary concerning the application of these extremely useful prefixes. First, note the difference between the symbols M and m. The ratio is one thousand million! Secondly, always convert a value into its basic unit before using it in an equation. If this is done, any unknown value in the equation can be found in terms of its basic unit.

Chapter 1

Basic electric units and circuits

1.1 SIMPLE ELECTRON THEORY

Electricity in the form of lightning must have been apparent to man from his earliest cave-dwelling days. The use of electricity has increased many times during the last 60 years, and with this increased use has come a greater awareness of the nature of electricity. Present-day theories are based on the theory of atomic structure, although our knowledge is still far from complete.

The Atom

All matter is composed of **atoms**, which often arrange themselves into groups called **molecules**. An atom is so very small that our minds are unable to appreciate what vast numbers of them make up even a very small piece of material. Eight million typical atoms, placed end to end, would stretch for about one millimetre.

The atom itself is not solid, but is composed of even smaller particles separated by space. At the centre of each atom is the **nucleus**, which is made up of various particles, including **protons**. These protons are said to have a **positive charge**. The **electrons**, which complete the atom, are in a constant state of motion, circling the nucleus in the same way as a satellite circles the Earth. Each electron has a **negative charge**. Atoms of different materials differ from one another by having different numbers of electrons, but, in the complete state, every atom has equal numbers of protons and electrons, so that positive and negative charges cancel out to leave the atom electrically neutral. The atoms in solids and liquids are much closer together than they are in gases, and in solids they are held in a definite pattern for a given material.

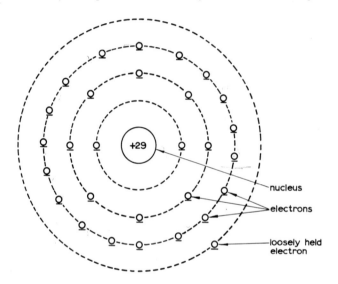

Fig. 1.1 Simplified representation of copper atom

Where there are more than two electrons in an atom, they arrange their paths of motion into **shells**. Fig. 1.1 shows a simple representation of a copper atom, which has 29 electrons and 29 protons. The electron in the outer shell is weakly held in position, and often breaks free, moving at random among the other copper atoms. An atom that has lost an electron in this way is left with an overall positive charge, since it has a positive proton in excess of those required to balance the effect of its negative electrons. Such an incomplete atom is often called a **positive ion**.

The movement of free electrons in a conductor depends on the laws of electric charge, which are

(a) like charges repel
(b) unlike charges attract.

Electric current

Fig. 1.2a represents a block of conducting material, containing free electrons moving at random among positive ions. If a battery is connected across the block as shown in Fig. 1.2b, free electrons close to the positive plate will be attracted to it, since unlike charges attract. Free electrons near the negative plate will be repelled from it, and a steady drift of electrons will take place through the material from the negative battery terminal to the positive battery terminal. For each electron entering the positive terminal, one will be ejected from the negative terminal, so that the number of electrons in the material remains constant.

Since the atoms that have become positive ions are unable to move in a solid, they do not drift to the negative terminal.

The rate of movement of electrons through the solid is very slow, but, since free electrons throughout the material start to drift immediately the battery is connected, there is very little delay in the demonstration of the effects that occur as a result of this movement.

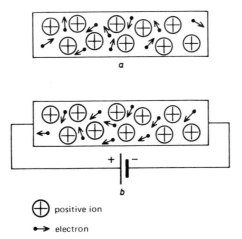

⊕ positive ion

•→ electron

Fig. 1.2 *a* Random movement of electrons in conductor,
b Drift of electrons towards positive plate when battery is connected

The drift of electrons *is* the electric current; so, to some extent, we have been able to answer the question 'What is electricity?' However, we have no clear understanding of the nature of an electron, so our knowledge is far from complete.

The electrons which enter the battery through the positive plate are passed through it, and are ejected from the negative plate into the conductor. Thus the electrons circulate, but must have a continuous conducting path, or **closed circuit**, in which to do so. If the circuit is broken, the drift of electrons will cease immediately.

Current direction
The knowledge that an electric current consists of a drift of electrons is of recent origin. Long before this theory was put forward, electric current was thought of as due to an 'electric fluid' which flowed in conductors from the positive plate of a battery to the negative. This direction of current, called **conventional current direction**, was thought to be correct for many years, so that many rules were based on it. We now know that this assumed current direction is incorrect and that current in a solid actually consists of an electron drift in the opposite direction. Despite this, we still continue, by convention, to indicate current direction, external to the source, as being from the positive to the negative terminal. In most of the applications we shall consider, the actual direction of current does not affect the performance of equipment; because of this, we shall continue to use conventional current direction.

The directions of electron drift and conventional current are shown in Fig. 1.3.

1.2 QUANTITY OF ELECTRICITY AND UNIT OF CURRENT

If we wish to measure a length we do so with a rule which is marked off in specific units of length. Since an electric current is invisible, we must use special instruments for measuring it, most of these instruments depending for their operation on the magnetic field set up by a current in a conductor. These instruments are described in Section 1.9.

The first unit we shall consider is the unit of electrical charge or quantity of electricity. It may seem that the electron could be used, but it is far too small for practical purposes. The unit used is the **coulomb** (symbol C), which is very much larger than the electron, the charge of over six million million million electrons equalling that of one coulomb. If a body has a surplus of electrons it is said to be **negatively**

charged, whereas, if it has a shortage of electrons, it is **positively charged**. Both these amounts of charge could be measured in coulombs.

A newcomer to the theory of electricity may be struck by the curious names applied to units. Most units are named after great scientists, like the unit of electrical charge, the coulomb, which is named after Charles Coulomb, a French physicist (1736–1806).

If the drift of electrons in a conductor takes place at the rate of one coulomb per second, the resulting current is said to be a current of one **ampere** (symbol **A**). Thus, a current of one ampere indicates that charge is being transferred along the conductor at the rate of one coulomb per second; hence

$$Q = It$$

where Q = charge transferred in coulombs
I = current in amperes
t = time during which the current flows in seconds

EXAMPLE 1.1

If a total charge of 500 C is to be transferred in 20 s, what current must flow?

$$Q = It, \text{ and thus } I = \frac{Q}{t}$$

Therefore

$$I = \frac{500}{20} \text{ amperes} = 25 \text{ A}$$

EXAMPLE 1.2

A current of 12·5 A passes for 2 min. What quantity of electricity is transferred?

$$Q = It$$
$$= 12·5 \times 2 \times 60 \text{ coulombs}$$
$$= 1500 \text{ C}$$

EXAMPLE 1.3

A current of 0·15 A must transfer a charge of 450 C. For how long must the current pass?

$$Q = It, \text{ so } t = \frac{Q}{I}$$

$$t = \frac{450}{0·15} \text{ seconds}$$

$$= 3000 \text{ s or } 50 \text{ min}$$

1.3 EFFECTS OF ELECTRIC CURRENT

Electrons are far too small to be seen even with the best available microscopes, and the detection of current would be impossible if it did not produce effects that are more easily detected. There are many such effects, but the three most important are **heating, chemical** and **magnetic**.

When current flows in a wire, heat is generated. The amount of heat produced in this way depends on a number of factors, which will be considered later, but can be controlled by the variation of current, of conductor material and of conductor dimensions. In this way, the conductor can be made red or white hot as with an electric fire or filament lamp, or can be made to carry current and remain reasonably cool as with an electric cable.

When current passes through chemical solutions, it can cause basic chemical changes to take place in them. Examples of this are the battery of cells, and electroplating. Some of these chemical effects will be further considered in Chapter 8.

A current flowing in a coil gives rise to a magnetic field, and this principle is the basis of many electrical devices such as the motor, the relay and the bell. The magnetic effect is the subject of Chapter 6. Fig. 1.3 shows a circuit in which the same current passes in turn through a filament lamp (heating effect), an electromagnet (magnetic effect) and a lead-acid cell (chemical effect). The heating and magnetic effects will be

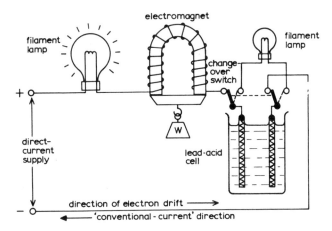

Fig. 1.3 Circuit to illustrate heating, magnetic and chemical effects of electric current

apparent owing to the heating of the lamp filament and the attraction of the iron armature. The chemical effect is demonstrated if the changeover switch is operated, when energy stored in the cell will cause the small filament lamp to glow.

Table 1 lists some of the common devices relying on these three electrical effects.

Table 1 Some devices relying on effects of electric current

Magnetic effect	Heating effect	Chemical effect
Relay	Filament lamp	Cells and batteries
Bell	Electric heater	Electroplating
Contactor	Electric cooker	Fuel cell
Telephone	Electric iron	
Motor	Electric kettle	
Generator	Television tube	
Transformer	Fuse	
Circuit breaker	Circuit breaker	
Electric clock	Welder	
Tape recorder	Furnace	
Ammeter		
Voltmeter		

1.4 ELECTRIC CONDUCTORS AND INSULATORS

Electrical conductors
It has been stated that an electric current is the drift of free electrons in a solid. It follows that, for a material to be capable of carrying current, the atoms of which it is composed must have loosely held electrons, which become detached at normal temperatures or can be detached by the application of an electric charge. Such materials are called **electrical conductors**. A list of conductors, with remarks on their properties and uses, is given in Table 2.

Silver is the best electrical conductor, but its high cost and poor physical properties prevent its use as a cable material. Copper is next in conducting properties to silver. Its malleability, and the ease with which it can be drawn into strands, makes it the natural choice as a conductor for cables; many heavy supply cables, and almost all wiring cables, have copper conductors.

Table 2 Electrical conductors

Material	Properties	Application
Aluminium	Low cost and weight	Power cables
Brass	Easily machined; resists corrosion	Terminals, plug pins
Carbon	Hard; low friction with other metals	Machine brushes
Chromium	Hard; resists corrosion	Heating elements (with nickel)
Copper	Very good conductor; soft and easily drawn into wires; capable of hardening	All cables, busbars
Gold	Expensive; does not corrode	Plating on contacts
Iron and steel	Common metal	Conduits, trunking, fuseboard cases etc. (protective conductor)
Lead	Does not corrode; bends easily	Cable sheaths (protective conductor)
Mercury	Liquid at normal temperatures; vaporises readily	Special contacts, discharge lamps
Nickel	Hard; resists corrosion	Heating elements (with chromium),
Silver	Expensive; the best conductor	Fine instrument wires, plating on contacts
Sodium	Vaporises readily	Discharge lamps
Tin	Resists attack by sulphur	Coating on copper cables insulated with vulcanised rubber
Tungsten	Easily drawn into fine wires	Lamp filaments

Aluminium is a poorer conductor than copper, but it is lighter. As copper prices have increased in recent years, aluminium prices have remained stable, so aluminium is a direct competitor to copper for power cables. Since aluminium is not as flexible as copper, cannot be drawn into such fine wires, and poses connection problems owing to rapid surface corrosion, the IEE Wiring Regulations forbid its use in the form of small wiring cables. Such cables, however, are made and used.

Electrical insulators

If a material is composed of atoms which have all their electrons tightly bound to them, there will be no free electrons available to form an electric current, and none can flow. Such materials are called **electrical insulators**. There are very many types of insulating material, but a few of those in common use in the electrical industry are listed in Table 3.

Table 3 Electrical insulators

Material	Properties	Application
Rubber / Flexible plastics	Flexible; life affected by high temperature	Cable insulation (small and medium sizes)
Impregnated paper / Varnished cambric	Rather stiff, but unaffected by moderate temperatures; hygroscopic	Cable insulation (medium and large sizes)
Magnesium oxide (mineral insulation)	Powder; requires containing sheath; not affected by very high temperatures; very hygroscopic	Mineral-insulated cables
Mica	Insulation not affected by high temperatures	Kettle elements, toaster elements etc.
Asbestos / Glass fibre	Reasonably flexible; not affected by high temperatures	Cable insulation in cookers, fires etc.
Porcelain	Hard and brittle; easily cleaned	Fuse carriers, overhead-line insulators etc.
Rigid plastics	Not so expensive as porcelain, and less brittle	Fuse carriers, switches sockets, plugs etc.

Table 3 is far from complete. For example, many new types of plastics have been developed as cable insulation, each having special properties. Polyvinyl chloride (p.v.c.) is the most used insulating and sheathing material for internal normal-temperature applications. Polychloroprene (p.c.p.) has particularly good weather-resisting properties, as has chlorosulphonated polyethelyne (c.s.p.), which has also increased resistance to physical damage. These are but three examples of the numerous insulating materials now available to the engineer and craftsman.

No material is a perfect insulator, and all will pass a small 'leakage current'. This leakage is almost always so small compared with the operating currents of the equipments that it may be ignored.

Conductors and insulators

An electric cable is a very good example of the application of conductors and insulators. Fig. 1.4 shows a typical twin housewiring cable with protective conductor, the copper current-carrying conductors being insulated with p.v.c. An overall sheath of p.v.c. keeps the conductors together, and protects them from damage and dampness. The conductors are made of copper, which is softened by annealing to make it flexible. The flexibility if sometimes further improved by using stranded, instead of solid, conductors.

Fig. 1.4 Twin-and-earth p.v.c.-insulated p.v.c.-sheathed wiring cable

Semiconductors

Semiconductors have electrical properties lying between those of conductors and insulators. They occupy a very important place in such devices as rectifiers, which will be considered in Chapter 15.

1.5 ELECTRICAL ENERGY, WORK AND POWER

Before we can go on to consider the electrical force that results in electron drift in a conductor, we must look at the units used for measuring work and power. Fuller consideration of electrical energy, work and power is given in Chapter 5.

Energy and work

Energy and work are interchangeable, energy being used to do work. Both are measured in terms of force and distance. The unit of force in the SI is the **newton** (N), and the unit of distance or length is the **metre** (m). If a force moves through a distance, work is done and energy is used.

$$\text{Energy used} = \text{work done} = \text{distance moved (m)} \times \text{force for movement (N)}$$

Since the distance moved is in metres and the force required for the movement is in newtons, the unit is the metre-newton, more commonly called the **joule** (J).

EXAMPLE 1.4

A force of 2000 N is required to lift a machine. How much work is done if the machine is lifted through 3 m?

$$\text{Work done} = \text{distance} \times \text{force}$$
$$= 3 \times 2000 \text{ joules}$$
$$= 6000 \text{ J}$$

This work is mechanical, but we shall see that work can also be electrical.

Power

Power is the rate of doing work or of using energy. For instance, an electrician can cut a hole in a steel plate using a hand drill or an electric drill. With both, the effective work done will be the same, but the electric drill will cut the hole more quickly because its power is greater.

It follows that

$$\text{power} = \frac{\text{work or energy}}{\text{time}}$$

The SI unit of power is the **watt (W)**, which is a rate of doing work of one joule per second.

$$\text{Watts} = \frac{\text{joules}}{\text{seconds}}$$

Similarly, if we know how much power is being used and the time for which it is used, we can find the total energy used.

$$\text{work or energy} = \text{power} \times \text{time}$$

$$\text{or joules} = \text{watts} \times \text{seconds}$$

EXAMPLE 1.5

A works truck requires a force of 180 N to move it. How much work is done if the truck is moved 20 m, and what average power is employed if the movement takes 40 s?

$$\text{Work} = \text{distance} \times \text{force}$$
$$= 20 \times 180 \text{ joules}$$
$$= 3600 \text{ J}$$

$$\text{Average power} = \frac{\text{work}}{\text{time}}$$
$$= \frac{3600}{40} \text{ watts}$$
$$= 90 \text{ W}$$

1.6 ELECTROMOTIVE FORCE AND POTENTIAL DIFFERENCE

When an electric current flows, energy is dissipated. Since energy cannot be created, it must be provided by the device used for circulating the current. This device may be chemical, such as a battery; or mechanical, such as a generator; or it may have one of a number of other forms. Many years ago, electricity was thought to be a fluid which circulated as the result of a force, and thus the term **electromotive force** (e.m.f.), symbol E, came into use. The e.m.f. is measured in terms of the number of joules of work necessary to move one coulomb of electricity round the circuit, and thus has the unit joules/coulomb. This unit is also referred to as the **volt** (symbol V) so that

$$1 \text{ volt} = 1 \text{ joule/coulomb}$$

EXAMPLE 1.6

A battery with an e.m.f. of 6 V gives a current of 5 A round a circuit for five minutes. How much energy is provided in this time?

$$\text{Total charge transferred } Q = It = 5 \times 5 \times 60 \text{ C} = 1500 \text{ C}$$

$$\text{Total energy supplied} = (\text{joules/coulomb}) \times \text{coulombs}$$
$$= \text{volts} \times \text{coulombs}$$
$$= 6 \times 1500 \text{ joules} = 9000 \text{ J}$$

In example 1.6, each coulomb of electricity contained six joules of potential energy on leaving the battery. This energy was dissipated on the journey round the circuit, so that the same coulomb would possess no energy on its return to the battery. The amount of energy expended by one coulomb in its passage between any two points in a circuit is known as the **potential difference** (p.d.) between those points, and is measured in joules/coulomb, or volts.

A convenient definition of the volt is therefore that it is equal to the difference in potential between two points if one joule of energy is required to transfer one coulomb of electricity between them.

EXAMPLE 1.7

How much electrical energy is converted into heat each minute by an immersion heater which takes 13 A from a 240 V supply?

$$\text{Energy given up by each coulomb} = 240 \text{ J}$$

$$\text{Quantity of electricity flow per minute} = It = 13 \times 60 \text{ coulombs} = 780 \text{ C}$$

$$\text{Therefore energy converted in one minute} = (\text{joules/coulomb}) \times \text{coulombs}$$

$$= 240 \times 780 \text{ joules} = 187\,200 \text{ J}$$

1.7 RESISTANCE: OHM'S LAW

For a metallic conductor which is kept at a constant temperature, it is found that the ratio

$$\frac{\text{potential difference across conductor (volts)}}{\text{resulting current in conductor (amperes)}}$$

is constant, and this ratio is known as the **resistance** (symbol R) of the conductor. This important relationship was first verified by Dr. G. S. Ohm, and is often referred to as 'Ohm's law'. The unit of resistance is the **ohm** (Greek symbol Ω, 'omega'). A conductor has a resistance of one ohm if the p.d. between its ends is one volt when it carries a current of one ampere. A device intended to have a resistance is called a **resistor**.

The relationship expressed by Ohm's law, which is of fundamental importance in electrical engineering, can be simply written as a formula

$$V = I \times R$$

The subject of the formula can be changed to give

$$I = \frac{V}{R} \text{ or } R = \frac{V}{I}$$

EXAMPLE 1.8

An electric heater used on a 240 V supply carries a current of 12 A. What is its resistance?

$$R = \frac{V}{I}$$

$$= \frac{240}{12} \text{ ohms}$$

$$= 20\,\Omega$$

EXAMPLE 1.9

The insulation resistance between two cables is two million ohms. What leakage current will flow if a p.d. of 400 V exists between them?

$$I = \frac{V}{R}$$

$$= \frac{400}{2\,000\,000} \text{ amperes}$$

$$= 0{\cdot}000\,2 \text{ A} \text{ or } 0{\cdot}2 \text{ mA}$$

EXAMPLE 1.10

What p.d. exists across an earth continuity conductor of resistance $1{\cdot}2\,\Omega$ when a current of 25 A flows through it?

$$V = IR$$

$$= 25 \times 1{\cdot}2 \text{ volts}$$

$$= 30 \text{ V}$$

1.8 ELECTRIC CIRCUIT

For an electromotive force to drive electrons, it must be applied to a closed circuit. In practice, the circuit is likely to consist of a piece of apparatus connected to the source of e.m.f. by means of cables, which complete the circuit. Fig. 1.5 shows a simple circuit consisting of a source of e.m.f., and a resistor. A switch, shown closed in Fig. 1.5a is included in the circuit. Since the circuit is completed, the e.m.f. of the supply will provide a current, its value depending on the voltage of the supply and on the resistance of the circuit.

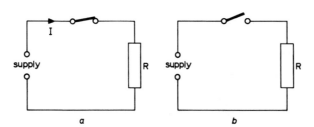

Fig. 1.5 a Closed circuit
b Open circuit

If the load has high resistance, the current will be small; if the load resistance is low, the current will be greater.

If the switch is opened, as in Fig. 1.5b, the gap between the opened contacts introduces a nearly infinite resistance into the circuit, so that the current falls to zero. We say that opening the switch has 'broken' the circuit.

Although there are some devices, including electronic valves and fluorescent lamps, which do not offer a complete metallic path for current, the majority of circuits are made up entirely of such conductors. If the conducting path is interrupted, the current ceases.

In circuits where high voltages are present, opening a switch may not break the circuit, the current continuing through the air between the contacts. The air carrying the current glows brightly, and gives off heat; it is called an **arc**. The majority of switches produce an arc when opened, but, in most cases, the arc disappears within a fraction of a second, and the circuit is broken before the heat from the arc can damage the switch or its surroundings.

1.9 AMMETERS AND VOLTMETERS

Although the presence of an electric current may produce effects which can be detected by the human senses, such effects are seldom useful as an indication of the value of the current. For instance, when a filament lamp glows, it is clearly carrying current, although we are unlikely to be able to judge the value. However, when the current in the lamp is reduced to about one-third of its normal value, the lamp ceases to glow.

Instruments for direct measurement of electric current are called **ammeters**, and will be considered in detail in the second volume. The principles of two types of instrument are discussed in Sections 7.8 and 11.6. Ammeters have low resistance, and are connected so that the current to be measured passes through them. Fig. 1.6 shows correct and incorrect ammeter connections. The ammeter will be damaged if incorrectly connected.

Fig. 1.6 makes it clear that the circuit symbol for an ammeter is a circle containing the letter A.

A **voltmeter** is another measuring instrument, used to indicate the potential difference between its two connections. To give an indication, the voltmeter must be connected across the device or circuit whose p.d. is to be indicated. Fig. 1.7 gives correct and incorrect voltmeter connections, and shows the voltmeter symbol as a circle containing the letter V. If connected incorrectly, the voltmeter is unlikely to be damaged, but, since it has high resistance, it will prevent the correct functioning of the circuit. The source of supply in the circuits of Figs. 1.6 and 1.7 is a battery of cells.

1.10 SERIES CIRCUITS

When a number of resistors are connected together end to end, so that there is only one path for current through them, they are said to be connected **in series**. An electrical appliance is connected in series with the cables feeding it; and, since the total current will depend on the resistance of the circuit as well as on the voltage applied to it, it is important to be able to calculate the resistance of the complete circuit if we know

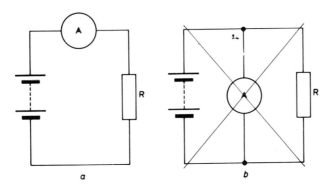

Fig. 1.6 Correct and incorrect connection of ammeter

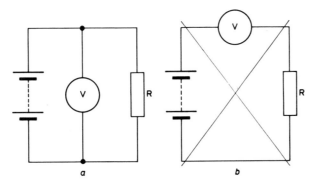

Fig. 1.7 Correct and incorrect connection of voltmeter

the values of the individual resistors connected in it. Fig. 1.8 shows three resistors, of values R_1, R_2 and R_3, respectively, connected in series across a supply of V volts. Let us assume that the resulting current is I amperes. If the total circuit resistance is R ohms,

$$R = \frac{V}{I}$$

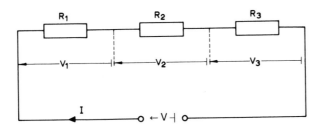

Fig. 1.8 Resistors connected in series

Now let the p.d. across each of the three resistors be V_1, V_2 and V_3 volts, respectively.

Then
$$V_1 = IR_1,\ V_2 = IR_2 \text{ and } V_3 = IR_3$$
but
$$V = V_1 + V_2 + V_3$$
$$= IR_1 + IR_2 + IR_3$$
$$= I(R_1 + R_2 + R_3)$$

12 Basic electric units and circuits

$$\frac{V}{I} = R_1 + R_2 + R_2$$

but

$$\frac{V}{I} = R, \text{ so that}$$

$$R = R_1 + R_2 + R_3$$

Thus the total resistance of any number of resistors connected together in series can be found by adding the values of the individual resistors, which must all be expressed in the same unit.

EXAMPLE 1.11

Resistors of 50 Ω and 70 Ω are connected in series to a 240 V supply. Calculate (*a*) the total resistance of the circuit, (*b*) the current, (*c*) the p.d. across each resistor.

(*a*) $$R = R_1 + R_2 = 50 + 70 \text{ ohms} = 120 \, \Omega$$

(*b*) $$I = \frac{V}{R} = \frac{240}{120} \text{amperes} = 2 \text{ A}$$

(*c*) $$V_1 = IR_1 = 2 \times 50 \text{ volts} = 100 \text{ V}$$

$$V_2 = IR_2 = 2 \times 70 \text{ volts} = 140 \text{ V}$$

Note: Supply voltage $V = V_1 + V_2 = 100 + 140 \text{ volts} = 240 \text{ V}$

Note that, for a series circuit,

(i) the same current flows in all resistors
(ii) the p.d. across each resistor is proportional to its resistance
(iii) the sum of the p.d.s across individual resistors is equal to the supply voltage.

EXAMPLE 1.12

An electric heater consists of an element of resistance 23·8 Ω and is fed from a 240 V supply by a 2-core cable of unknown resistance. If the current is 10 A, calculate the total resistance of the cable.

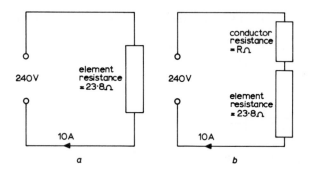

Fig. 1.9   Circuit diagrams for example 1.12

Fig. 1.9*a* shows the heater connected to the supply through a 2-conductor cable. Fig. 1.9*b* is an equivalent circuit in which conductors are assumed to have no resistance, the actual resistance of the conductors being replaced by the resistor R. There are two methods of solution.

Method 1

Total resistance $= \frac{V}{I} = \frac{240}{10} \text{ohms} = 24 \, \Omega$

But total resistance = element resistance + conductor resistance
Therefore conductor resistance = total resistance − element resistance
Conductor resistance = 24 − 23·8 ohms = 0·2 Ω

Method 2

P.D. across element = current × element resistance
$$= 10 \times 23\cdot 8 \text{ volts} = 238 \text{ V}$$
Supply voltage = p.d. across element + p.d. across conductors
Therefore p.d. across conductors = supply voltage − p.d. across element
$$= 240 - 238 \text{ volts} = 2 \text{ V}$$
Conductor resistance = $\dfrac{\text{conductor p.d.}}{\text{current}}$
$$= \frac{2}{10} \text{ ohms} = 0\cdot 2 \, \Omega$$

1.11 PARALLEL CIRCUITS

When each one of a number of resistors is connected between the same two points, they are said to be connected **in parallel**. In this form of connection, the total current divides, part of it flowing in each resistor. Since all the resistors are connected across the same two points, the p.d. across each one is the same. Fig. 1.10 shows resistors of values R_1, R_2 and R_3, respectively, connected in parallel to a supply of V volts. The currents in the resistors are I_1, I_2 and I_3, respectively, and the total current is I.

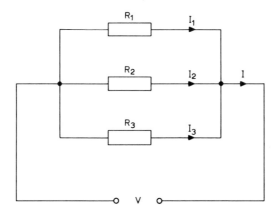

Fig. 1.10 Resistors connected in parallel

The total current divides itself among the resistors, so that
$$I = I_1 + I_2 + I_3$$
But, from Ohm's law,
$$I_1 = \frac{V}{R_1}, \; I_2 = \frac{V}{R_2}, \text{ and } I_3 = \frac{V}{R_3}$$
$$I = \frac{V}{R_1} + \frac{V}{R_2} + \frac{V}{R_3} = V\left(\frac{1}{R_1} + \frac{1}{R_2} + \frac{1}{R_3}\right)$$
$$\frac{I}{V} = \frac{1}{R_1} + \frac{1}{R_2} + \frac{1}{R_3}$$

If the equivalent resistance of the parallel circuit is R, $R = V/I$.

Therefore
$$\frac{I}{V} = \frac{1}{R} = \frac{1}{R_1} + \frac{1}{R_2} + \frac{1}{R_3}$$

The value $1/R$ is called the **reciprocal** of R. We can thus sum up the expression by saying that the reciprocal of the equivalent resistance of a parallel circuit is equal to the sum of the reciprocals of the resistances of the individual resistors.

Note that, for a parallel circuit,

(*a*) the same p.d. occurs across all resistors

14 Basic electric units and circuits

(b) the current in each resistor is inversely proportional to its resistance
(c) the sum of the currents in the individual resistors is equal to the supply current.

As for the series circuit, the resistances must all be expressed in the same unit before using them in the formula.

EXAMPLE 1.13

Calculate the equivalent resistance of four parallel-connected resistors of 6 Ω, 30 Ω, 5 Ω and 10 Ω, respectively.

$$\frac{1}{R} = \frac{1}{R_1} + \frac{1}{R_2} + \frac{1}{R_3} + \frac{1}{R_4}$$

$$\frac{1}{R} = \frac{1}{6} + \frac{1}{30} + \frac{1}{5} + \frac{1}{10} = \frac{5 + 1 + 6 + 3}{30} = \frac{15}{30}$$

$$R = \frac{1}{1/R} = \frac{30}{15} \text{ohms} = 2\,\Omega$$

It can be seen from this result that the equivalent resistance of any group of parallel-connected resistors is lower than that of the lowest-valued resistor in the group. If a number of equal-value resistors are connected in parallel, the equivalent resistance will be the value of one resistor divided by the number of resistors. For example, ten 10 Ω resistors in parallel have an equivalent resistance of 1 Ω.

EXAMPLE 1.14

Resistors of 16 Ω, 24 Ω and 48 Ω, respectively, are connected in parallel to a 240 V supply. Calculate the total current.

There are two ways of solving this problem.

Method 1

Find the equivalent resistance and use it with the supply voltage to find the total current.

$$\frac{1}{R} = \frac{1}{16} + \frac{1}{24} + \frac{1}{48} = \frac{3 + 2 + 1}{48} = \frac{6}{48}$$

$$R = \frac{48}{6} \text{ohms} = 8\,\Omega$$

$$I = \frac{V}{R} = \frac{240}{8} \text{amperes} = 30\,\text{A}$$

Method 2

Find the current in each resistor. Add these currents to give the total current.

$$\text{Current in 16 } \Omega \text{ resistor} = \frac{V}{R_{16}} = \frac{240}{16} \text{amperes} = 15\,\text{A}$$

$$\text{Current in 24 } \Omega \text{ resistor} = \frac{V}{R_{24}} = \frac{240}{24} \text{amperes} = 10\,\text{A}$$

$$\text{Current in 48 } \Omega \text{ resistor} = \frac{V}{R_{48}} = \frac{240}{48} \text{amperes} = 5\,\text{A}$$

$$\text{Total current} = 15 + 10 + 5 \text{ amperes} = 30\,\text{A}$$

There are many applications of parallel circuits. The elements of a 2-bar fire are connected in parallel, and the heat output is varied by switching one bar on or off as required. The two circuits in a cooker grill can be connected in three ways to give 3-heat control, as indicated in the following example.

Basic electric units and circuits 15

EXAMPLE 1.15

The grill of an electric cooker has two identical elements, each of resistance 48 Ω, which are connected in parallel for 'high' heat and in series for 'low' heat. One element only is used for 'medium' heat. Calculate the current drawn from a 240 V supply for each switch position.

Low The elements are in series (Fig. 1.11a)
Total resistance = 48 + 48 ohms = 96 Ω
Therefore $I = \dfrac{V}{R} = \dfrac{240}{96}$ amperes = 2·5 A

Medium One element only in use (Fig. 1.11b)
Therefore $I = \dfrac{V}{R} = \dfrac{240}{48}$ amperes = 5 A

High The elements are in parallel (Fig. 1.11c)
Total resistance = $\dfrac{48}{2}$ ohms = 24 Ω
Therefore $I = \dfrac{V}{R} = \dfrac{240}{24}$ amperes = 10 A

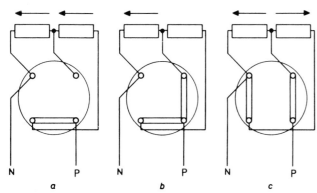

Fig. 1.11 3-heat switching circuits for example 1.15
(a) Low
(b) Medium
(c) High

If only two resistors are connected in series, there is a simpler way of calculating the equivalent resistance than by adding the reciprocals. It is particularly useful where resistor values are not whole numbers, and when therefore the lowest common denominator is not easily found. It can be shown that for two resistors in parallel,

$$R = \dfrac{R_1 \times R_2}{R_1 + R_2}$$

This method is often referred to as "product over sum". It should be noticed that it is only useful for TWO resistors. Although there are equivalent formulas for more resistors, they are so complicated as to be difficult to use. However, if there *are* more than two resistors, they can be considered in stages.

For example, consider that we need to find the equivalent resistance of three resistors, of 15 Ω, 10 Ω and 4 Ω all connected in parallel. Take the first two resistors and find their equivalent value, which we will call R_A.

$$R_A = \dfrac{R_1 \times R_2}{R_1 + R_2}$$

$$= \dfrac{10 \times 15}{10 + 15} \text{ ohm}$$

$$= \dfrac{150}{25} \text{ ohm}$$

$$= 6 \Omega$$

Now take this equivalent 6Ω resistor and put it in parallel with the third, 4Ω resistor.

$$R = \frac{R_A \times R_3}{R_A + R_3}$$

$$= \frac{6 \times 4}{6 + 4} \text{ ohm}$$

$$= \frac{24}{10} \text{ ohm}$$

$$= 2\cdot 4 \, \Omega$$

1.12 SERIES – PARALLEL CIRCUITS

A series-parallel circuit is one which is made up of series and parallel parts in combination. The possible number of combinations is endless, but all these circuits can be solved by simplification. A number of resistors, seen to be in series or in parallel, are replaced by one resistor which has the same effect on the circuit. This principle is explained in example 1.16.

EXAMPLE 1.16

Two banks of resistors are connected in series. The first bank consists of two resistors of 10 Ω and 40 Ω in parallel, and the second consists of three resistors, each of 12 Ω, connected in parallel. What is the resistance of the combination, and what current will be taken from a 12 V supply to which it is connected?

The circuit diagram is shown in Fig. 1.12a. To solve, we must look for groups of resistors connected in series or in parallel. The first bank consists of two resistors in parallel, so we must find the resistance of this combination.

$$\frac{1}{R} = \frac{1}{10} + \frac{1}{40} = \frac{4+1}{40} = \frac{5}{40}$$

Therefore $$R = \frac{40}{5} \text{ ohms} = 8 \, \Omega$$

Thus, the first group of resistors can be replaced by a single resistor of 8 Ω.

The second bank consists of three 12 Ω resistors in parallel. Its resistance can be found thus:

$$\frac{1}{R} = \frac{1}{12} + \frac{1}{12} + \frac{1}{12} = \frac{3}{12}$$

Therefore $$R = \frac{12}{3} \text{ ohms} = 4 \, \Omega$$

The second group can thus be replaced by a single 4 Ω resistor. Fig. 1.12b shows a simple series circuit which is the equivalent of that in Fig. 1.12a. These two circuits are not the same, but they have the same resistance and will take the same current when connected to a supply.

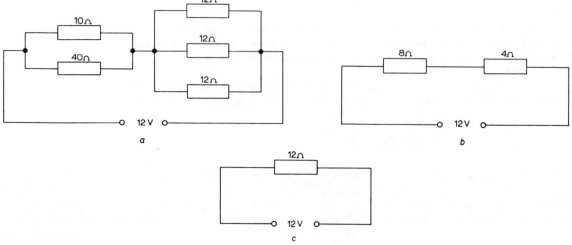

Fig. 1.12   Circuit diagrams for example 1.16

These two resistors can now be combined to a single equivalent resistor.

$$R = 8 + 4 \text{ ohms} = 12 \, \Omega$$

This is the resistance of the complete circuit. To find the current, Ohm's law is applied.

$$I = \frac{V}{R} = \frac{12}{12} \text{ ampere} = 1 \text{ A}$$

EXAMPLE 1.17

Fig. 1.13 shows a resistor network connected to a 240 V supply. Ammeters and voltmeters are to be connected to measure the current in each resistor, and the p.d. across each resistor. Redraw the diagram, adding these instruments. Calculate their readings.

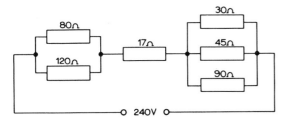

Fig. 1.13 Circuit for example 1.17

Ammeters must be connected so that the current to be measured passes through them. An ammeter must therefore be connected in series with each resistor.

Voltmeters must be connected so that the potential difference to be measured is also across their terminals. A voltmeter could thus be connected in parallel with each resistor. However, resistors connected in parallel clearly have the same potential difference across them, so one voltmeter connected across such a parallel group will suffice.

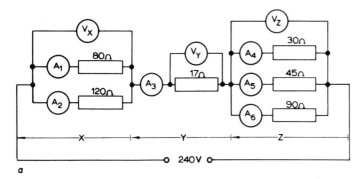

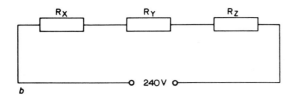

Fig. 1.14 Circuits for solution of example 1.17

Fig. 1.14a shows the circuit with ammeters and voltmeters added. It should be noticed that voltmeter V_x is connected across the 80 Ω resistor *and* ammeter A_1, as well as across the 120 Ω resistor *and* ammeter A_2. In fact, each of the ammeters has resistance and hence a potential difference appears across it when carrying current. Thus the voltmeter will read the sum of the p.d.s across a resistor and an ammeter. In practice, the resistance of an ammeter is so small that, in most cases, its potential difference can be ignored.

To find the readings on the instruments, we must calculate the p.d. across each resistor, as well as the current through each resistor.

The first step is to find the resistance of the whole circuit, and hence the total current. It is most important when dealing with circuits of this sort to be quite clear which part of the circuit is being considered. To this end, the three series-connected sections of the circuit have been called X, Y and Z, respectively (Fig. 1.14a). As in the previous example, the circuit must now be reduced to the simple series circuit shown in Fig. 1.14b.

$$\frac{1}{R_x} = \frac{1}{80} + \frac{1}{120} = \frac{3+2}{240} = \frac{5}{240}$$

Therefore
$$R_x = \frac{240}{5} \text{ ohms} = 48\,\Omega$$

$$R_y = 17\,\Omega$$

$$\frac{1}{R_z} = \frac{1}{30} + \frac{1}{45} + \frac{1}{90} = \frac{3+2+1}{90} = \frac{6}{90}$$

Therefore
$$R_z = \frac{90}{6} \text{ ohms} = 15\,\Omega$$

$$\text{Total circuit resistance} = R_x + R_y + R_z$$
$$= 48 + 17 + 15 \text{ ohms}$$
$$= 80\,\Omega$$

$$\text{Current from supply, } I = \frac{V}{R} = \frac{240}{80} \text{ amperes} = 3\,\text{A}$$

When this current flows in the circuit of Fig. 1.14b, the voltage drop across R_x,

$$V_x = IR_x = 3 \times 48 \text{ volts} = 144\,\text{V}$$

Since voltmeter V_x is connected across a group of resistors having the same effect as R_x, V_x will read 144 V.

Similarly,
$$V_y = IR_y = 3 \times 17 = 51\,\text{V}$$

and voltmeter V_y will read 51 V

Similarly,
$$V_z = IR_z = 3 \times 15 \text{ volts} = 45\,\text{V}$$

and voltmeter V_z will read 45 V

Note: As pointed out in Section 1.10, for a series circuit, the sum of the p.d.s across the individual resistors is equal to the supply voltage.

To check this,
$$144\,\text{V} + 51\,\text{V} + 45\,\text{V} = 240\,\text{V}$$

We can now calculate the current in each resistor.

The resistors in section X each have the section p.d. of 144 V applied to them.

Thus
$$I_1 = \frac{V_x}{R_1} = \frac{144}{80} \text{ amperes} = 1\cdot8\,\text{A}$$

so, ammeter A_1 reads 1·8 A

$$I_2 = \frac{V_x}{R_2} = \frac{144}{120} \text{ amperes} = 1\cdot2\,\text{A}$$

and ammeter A_2 reads 1·2 A

Note: As pointed out in Section 1.11, for a parallel circuit, the sum of the currents in individual resistors is equal to the supply current. To check this,

$$1\cdot8\,\text{A} + 1\cdot2\,\text{A} = 3\,\text{A}$$

The 17 Ω resistor in section Y has 51 V applied to it.

$$I_3 = \frac{V_y}{R_3} = \frac{51}{17} \text{ amperes} = 3\,\text{A}$$

Thus ammeter A_3 reads 3 A.

Inspection of the circuit would have confirmed this current without calculation. Since the 17 Ω resistor has no other resistor in parallel with it, it must carry the whole of the circuit current.

The resistors in section Z each have 45 V applied to them.

Thus
$$I_4 = \frac{V_z}{R_4} = \frac{45}{30} \text{ amperes} = 1\cdot5 \text{ A}$$

and ammeter A_4 reads 1·5 A

$$I_5 = \frac{V_z}{R_5} = \frac{45}{45} \text{ amperes} = 1 \text{ A}$$

and ammeter A_5 reads 1 A

$$I_6 = \frac{V_z}{R_6} = \frac{45}{90} \text{ amperes} = 0\cdot5 \text{ A}$$

and ammeter A_6 reads 0·5 A

Check:
$$1\cdot5 \text{ A} + 1\cdot0 \text{ A} + 0\cdot5 \text{ A} = 3 \text{ A}$$

EXAMPLE 1.18

The circuit shown in Fig. 1.15 takes a current of 6 A from the 50 V supply. Calculate the value of the resistor R_4.

There are several ways of tackling this problem. One of the simplest demonstrates a different approach from those possible in the previous worked examples.

First, mark the currents on the diagram (this has already been done in Fig. 1.15).

R_1 has the 50 V supply directly across it, so
$$I_1 = \frac{V}{R_1} = \frac{50}{25} \text{ amperes} = 2 \text{ A}$$

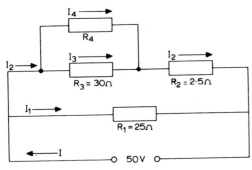

Fig. 1.15 Diagram for example 1.18

Since the total current is 6 A, it follows that the current in the upper part of the circuit, I_2, must be equal to 6 A minus 2 A, or 4 A. The p.d. across R_2, $V_2 = I_2 R_2 = 4 \times 2\cdot5 \text{ V} = 10 \text{ V}$.

The parallel combination of R_3 and R_4 is in series with R_2 across the 50 V supply. Since the p.d. across R_2 is 10 V, the p.d. across both R_3 and R_4 must be 50 V minus 10 V, or 40 V.

Thus
$$I_3 = \frac{V_3}{R_3} = \frac{40}{30} \text{ amperes} = 1\cdot33 \text{ A}$$

The currents I_3 and I_4 must sum to the current I_2, so
$$I_4 = I_2 - I_3 = 4 - 1\cdot33 \text{ amperes} = 2\cdot67 \text{ A}$$

Applying Ohm's law to R_4,
$$R_4 = \frac{V_4}{I_4} = \frac{40}{2\cdot67} \text{ ohms} = 15 \, \Omega$$

Note that in the working above, $1\tfrac{1}{3}$ A would have been exact whereas 1·33 A is correct to only two decimal places. In general, however, we avoid the use of fractions in electrical calculations, because instruments are invariably scaled decimally. Thus an ammeter could perhaps be read as 1·33 A but not as $1\tfrac{1}{3}$ A.

There are many circuits, often called networks, which are not series-parallel connections and to which the methods indicated will not apply. These must be solved by more advanced circuit theorems, which are beyond our scope at this stage.

However, the vast majority of the circuits encountered by the electrical craftsman are of the series, parallel or series-parallel type.

1.13 SUMMARY OF FORMULAS FOR CHAPTER 1

$$Q = It \qquad I = \frac{Q}{t} \qquad t = \frac{Q}{I}$$

where
Q = electrical charge or quantity, C
I = current, A
t = duration of current, s

$$W = dF \qquad d = \frac{W}{F} \qquad F = \frac{W}{d}$$

where
W = energy used or work done, mN or J
F = force applied, N
d = distance moved, m

$$P = \frac{W}{t} \qquad W = Pt \qquad t = \frac{W}{P}$$

where
P = power or rate of doing work, W

$$W = VQ \qquad Q = \frac{W}{V} \qquad V = \frac{W}{Q}$$

where
V = potential difference, V

For resistors in series,

$$R = R_1 + R_2 + R_3 \ldots$$

For resistors in parallel,

$$\frac{1}{R} = \frac{1}{R_1} + \frac{1}{R_2} + \frac{1}{R_3} + \ldots$$

where
R = total circuit resistance, Ω
R_1, R_2, R_3 etc. = individual circuit resistances, Ω

1.14 EXERCISES

1. A current of 10 A flows for two minutes. What charge is transferred?

2. For how long must a current of 4 mA flow so as to transfer a charge of 24 C?

3. What current must flow if 100 C is to be transferred in 8 seconds?

4. (a) An electron is situated between a positive and a negative charge. Towards which charge will the electron move?
   (b) 224 coulombs of electricity pass a given point in a circuit in 1 min 10 s. Calculate the current in the circuit. (NCTEC)

5. Briefly describe any one application of the chemical effect of an electric current. (NCTEC)

6. Name three materials used as electrical conductors, and three which are used as electrical insulators. (NCTEC)

7. A force of 450 N is required to lift a bundle of conduit. How much work is done if it is raised from the floor to the roof rack of a van 2 m high?

8. Energy of 2 J is required to close a contactor against a spring exerting a force of 80 N. How far does the contactor move?

9. When an electric motor is pushed 30 m across a level floor, 4500 J or work is done. What force is needed to move the motor?

10. A force of 35 N is required at the end of a spanner 0·2 m long to move a nut on a thread. How much work is done in giving the nut one complete turn?

11. A d.c. generator has an e.m.f. of 200 V and provides a current of 10 A. How much energy does it provide each minute?

12. A photocell causes a current of 4 µA in its associated circuit, and would take 1000 days to dissipate an energy of 1 mJ. What e.m.f. does it provide?

13. An electric blanket is required to provide heat energy at the rate of 7200 J/min from a 240 V supply. What current will flow?

14. If the total resistance of an earth fault loop is 4 Ω, what current will flow in the event of a phase-to-earth fault from 240 V mains?

Basic electric units and circuits   21

15   During a flash test, a voltage of 20 kV is applied to a cable with an insulation resistance of 5 MΩ. What will be the leakage current?

16   What is the resistance of an immersion heater element that takes 12·0 A from a 240 V supply?

17   An indicator lamp has a hot resistance of 50 Ω and a rated current of 0·2 A. What is its rated voltage?

18   The p.d. across an earth-continuity conductor of resistance 0·3 Ω is found to be 4·5 V. What current is the conductor carrying?

19   Four resistors of values 5 Ω, 15 Ω, 20 Ω and 40 Ω, respectively, are connected in series to a 240 V supply. Calculate the resulting current and the p.d. across each resistor.

20   A 6 Ω resistor and a resistor of unknown value are connected in series to a 12 V supply, when the p.d. across the 6 Ω resistor is measured as 9 V. What is the value of the unknown resistor?

21   Calculate the resistance of the element of a soldering iron that takes ½ A from 240 V mains when connected to them by cables having a total resistance of 0·2 Ω.

22   Calculate the equivalent resistance of each of the following parallel-connected resistor banks:
   (a)   2 Ω and 6 Ω;
   (b)   12 MΩ, 6 MΩ and 36 MΩ; and
   (c)   100 μΩ, 600 μΩ and 0·0012 Ω.

23   Answer the following questions by writing down the missing word or words:
   (a)   A fuse protects a circuit against . . . . and uses the . . . . effect of an electric current.
   (b)   The unit of quantity of electricity is called the . . . .
   (c)   Two good conductors of electrical current are . . . . and . . . .
   (d)   Two items of electrical equipment that use the electromagnetic effect are . . . . and . . . .
   (e)   Two insulating materials used in the electrical industry are . . . . and . . . .
   (f)   In an electrical circuit, the electron flow is from the . . . . terminal to the . . . . terminal.
   (g)   A bimetallic strip uses the . . . . effect of an electric current.
   (h)   The electron has a . . . . charge.
   (i)   An e.m.f. of 72 V is applied to a circuit, and a current of 12 A flows. The resistance of the circuit is . . . .
   (j)   The effective resistance of two 10 Ω resistors connected in parallel is . . . .
   (k)   Quantity of electricity = ... × . . . .
   (l)   0·36 amperes = ... milliamperes.
   (m)   3·3 kilovolts = ... volts. (ULCI)

24   Three resistors are connected in parallel across a supply of unknown voltage. Resistor A is of 7·5 Ω and carries a current of 4 A. Resistor B is of 10 Ω, and resistor C is of unknown value but carries a current of 10 A. Calculate the supply voltage, the current in resistor B, and the value of resistor C.

25   Three parallel-connected busbars have resistances of 0·1 Ω, 0·3 Ω and 0·6 Ω, respectively, and, in the event of a short-circuit, would be connected directly across a 400 V supply. Calculate the equivalent resistance of the combination, the total fault current, and the current in each busbar.

26   Resistors of 7 Ω, 14 Ω and 21 Ω, respectively, are connected in parallel. This bank is connected in series with a 2·5 Ω resistor across a supply of unknown potential. The current flow in the 2·5 Ω resistor is 2 A. What is the supply voltage?

27   (a)   Show, by separate drawings for 'high', 'medium' and 'low' heat positions, the connections of a series-parallel switch controlling two separate sections of resistance wire forming the element of a heating appliance.
   (b)   If the two sections of resistance wire are of equal resistance, what is the proportional current flow and heating effect in the 'medium' and 'low' positions relative to the 'high' position?

28   Three resistors, having resistances of 4·8 Ω, 8 Ω and 12 Ω, all connected in parallel, are supplied from a 48 V supply. Calculate the current through each resistor and the current taken from the supply.
Calculate the effective resistance of the group. (ULCI)

29   Two resistances of 4 Ω and 12 Ω, respectively, are connected in parallel with each other. A further resistance of 10 Ω is connected in series with the combination. Calculate the respective direct voltages which should be applied across the whole circuit,
   (a)   to pass 6 A through the 10 Ω resistance
   (b)   to pass 6 A through the 12 Ω resistance.

30   (a)   State Ohm's law in words and by symbols.
   (b)   Three relay coils, each of 7·5 Ω resistance, are connected in parallel. What current would flow in the circuit under a pressure of 12 V?
   (c)   Calculate the resistance required to limit the current in a circuit to 25 A with an applied pressure of 210 V.

31 Three banks of resistors are connected in series across a 240 V supply. Bank A consists of three resistors, R_1, R_2 and R_3, each of resistance 60 Ω, connected in parallel. Bank B comprises two resistors, R_4 of resistance 40 Ω and R_5 of resistance 120 Ω, connected in parallel. Bank C has three parallel-connected resistors, $R_6 = 50\,\Omega$, $R_7 = 100\,\Omega$ and $R_8 = 300\,\Omega$. Calculate
(a) the resistance of the complete circuit
(b) the current in each resistor
(c) the p.d. across each resistor.

32 A resistance network is connected as shown in Fig. 1.16, and takes a total current of 2·4 A from the 24 V supply. Calculate the value of the resistor R_x.

33 The voltmeter shown in Fig. 1.17 reads 39 V. What is the value of the resistor across which the voltmeter is connected?

34 Three banks of resistors are connected in series. Bank 1 comprises 40 Ω and 24 Ω in parallel; bank 2 comprises a single resistor of 5 Ω; bank 3 comprises four resistors in parallel, each of 16 Ω.

Ammeters are inserted to measure the current through each resistor, and voltmeters to measure the voltage across each bank. Sketch a layout of the apparatus, and calculate the reading on each instrument if 48 V be applied to the circuit. (C & G)

35 A bank of three paralleled resistors each of 12 Ω is connected in series with another bank consisting of three resistors in parallel, their values being 4 Ω, 4 Ω and 2 Ω. If a 100 V supply is applied across this circuit, what current would flow through the 2 Ω resistor?

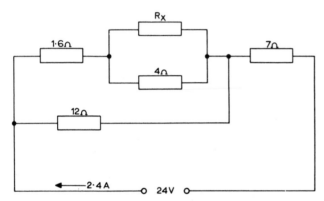

Fig. 1.16   Circuit diagram for exercise 1.32

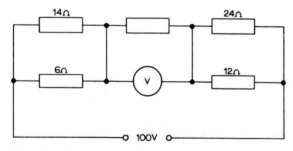

Fig. 1.17   Circuit diagram for exercise 1.33

Chapter 2
Resistance and resistors

2.1 INTRODUCTION

The conductors used to connect together the components of a circuit have the property of **resistance**. Usually, this resistance will be so small, in comparison with that of the components themselves, that it may be neglected for the purposes of circuit calculations. The conductor resistance will be present, however, and, sometimes, as indicated in Section 2.7, the volt drop in resistive conductors must be taken into account.

If conductors without any resistance at all were possible at a reasonable cost, they would be used widely.

Probably the most common circuit component is the **resistor**. Such a component is deliberately made to offer resistance to electric current. Resistors can exist in a very wide variety of shapes and sizes, a few of the most common constructions being shown in Fig. 2.1 The British Standard objective symbol for a **fixed resistor** is shown in Fig. 2.2a, with the alternative symbol in Fig. 2.2b. The new symbol has been in use since 1966. The alternative symbol is still in very wide use, and is likely to be so for many years.

a

b

Fig. 2.1 Fixed resistors
 a Carbon low-power resistor
 b Wire-wound encapsulated resistor

In many practical circuits, as well as in laboratory work, it is often necessary to use a **variable resistor**. This is usually a resistor with a sliding contact, so that the resistance between one end and the slider can be adjusted by movement of the latter. Some examples of variable resistors are shown in Fig. 2.3. Fig. 2.4 shows the objective and alternative general symbols for a variable resistor, and Fig. 2.5 shows the symbols that are used when the connection to the slider must be shown. A variable resistor is sometimes called a **rheostat**.

Many variable resistors can be connected as voltage dividers or **potential dividers** (Fig. 2.6). The voltage V_1 that appears at the output terminals will depend on the setting of the slider, and on the current taken from the output. Provided that the output current is very small, it follows that

$$V_1 = V \times \frac{R_1}{R}$$

This method of connection is often used to provide a variable-voltage supply.

2.2 EFFECT OF DIMENSIONS ON RESISTANCE

Let us imagine a cube of conducting material (Fig. 2.7a) that has a resistance of r ohms between opposite faces. If, say, five of these cubes are joined (Fig. 2.7b), they form a series resistor chain which has a total resistance of $5r$ ohms. If, say, four conductors, each made up of five cubes, are placed side by side (Fig. 2.7c), the result will be one conductor having four times the cross-sectional area of one of the original conductors.

c

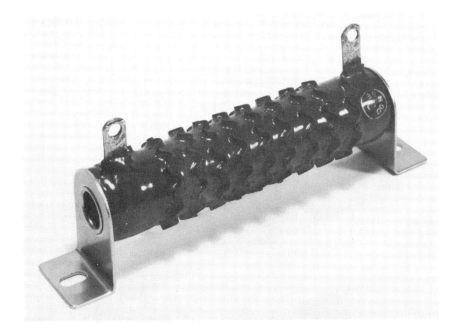

d

Fig. 2.1 Fixed resistors
    c Vitreous enamelled wire-wound resistor
    d High-power asbestos woven-mat resistor

The total resistance R of this composite conductor will be that of four 5r ohm resistors connected in parallel:

$$\frac{1}{R} = \frac{1}{5r} + \frac{1}{5r} + \frac{1}{5r} + \frac{1}{5r} = \frac{4}{5r}$$

Therefore
$$R = \frac{5r}{4} \text{ ohms}$$

This is the resistance of a conductor five cubes long and of four cubes in cross-section.

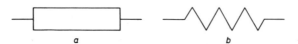

Fig. 2.2 Circuit symbols for fixed resistor
        a Objective symbol   *b* Alternative symbol

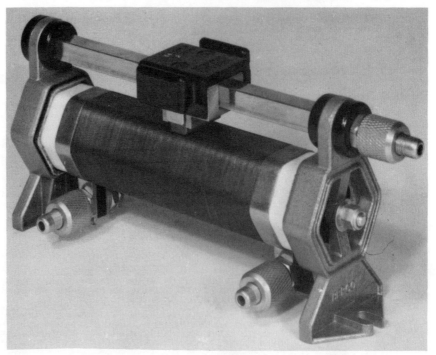

Fig. 2.3 Variable resistors
        a Wire-wound resistor with adjustable tapping
        b Wire-wound resistor with continuous variation

If the conductor had a length l and a uniform cross-sectional area a, its resistance would be

$$R = \frac{r \times l}{a}$$

In this way, the resistance of a conductor can be expressed in terms of its length, its cross-sectional area and the resistance between opposite faces of a cube of material identical to that of which the conductor is made.

Conductors are not often shaped so that they can be considered directly as a number of cubes, but the result obtained is true for a cable of any shape if its cross-sectional area is uniform throughout its length.

Fig. 2.3 Variable resistors
  c Wire-wound toroidal potential divider

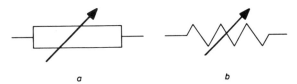

Fig. 2.4 Circuit symbols for variable resistor
  a Objective symbol   b Alternative symbol

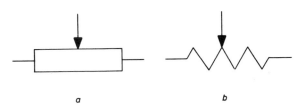

Fig. 2.5 Circuit symbols for resistor with moving contact
  a Objective symbol   b Alternative symbol

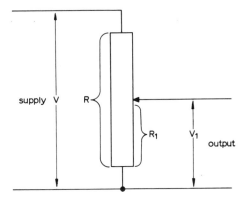

Fig. 2.6 Variable voltage output from potential divider

28 Resistance and resistors

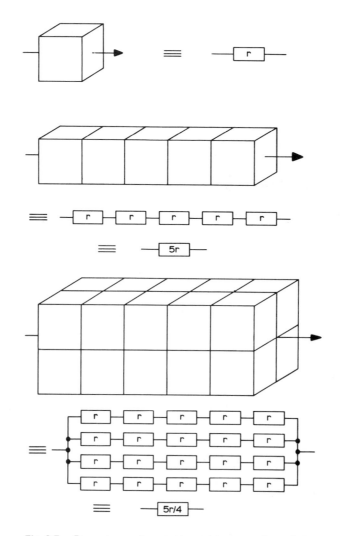

Fig. 2.7 Dependence of conductor resistance on dimensions

EXAMPLE 2.1

The resistance of 50 m of a certain cable is 0·15 Ω. Calculate the resistance of 800 m of this cable.

From the information above, $R \propto \dfrac{l}{a}$.

For a cable of a given size, the cross-sectional area a is constant, so resistance is directly proportional to length, or $R \propto l$. 800 m of cable is 800/50, or 16 times, longer than 50 m of cable. Thus, the resistance of 800 m of cable will be 16 times that of 50 m of cable.

Therefore $$R = 16 \times 0·15 \text{ ohms} = 2·4 \, \Omega$$

EXAMPLE 2.2

If the resistance per 100 m of a cable of cross-sectional area 2 mm² is 0·075 Ω, what is the resistance of 100 m of a cable made of the same conductor material which has a cross-sectional area of 5 mm²?

Since l is constant, $R \propto 1/a$

5 mm² is 5/2 or 2·5 times larger than 2 mm².

Since resistance is *inversely* proportional to cross-sectional area, the resistance of the second cable will be 2·5 times *smaller* than that of the first cable, or

$$R = \frac{0·075}{2·5} \text{ ohms} = 0·03 \, \Omega$$

EXAMPLE 2.3

If the resistance of 100 m of a 4 mm² cable is 0·06 Ω, what will be the resistance of 700 m of 7 mm² cable made of the same conductor material?

$$R \propto \frac{l}{a}$$

Therefore
$$R_2 = R_1 \times \frac{l_2}{l_1} \times \frac{a_1}{a_2}$$

where subscripts 1 and 2 refer to the first and second cables, respectively.

Therefore
$$R_2 = 0{\cdot}06 \times \frac{700}{100} \times \frac{4}{7} \text{ ohms}$$

$$= 0{\cdot}24 \, \Omega$$

EXAMPLE 2.4

In the manufacture of aluminium cable, a 10 m length of thick circular rod, which has a resistance of 0·015 Ω, is drawn out until its new diameter is one-fifth of its previous measurement. Calculate the length of the drawn cable and its resistance.

By drawing the aluminium bar through a die, its length is increased by reducing its diameter. The volume must, however, be the same before and after the drawing operation, since no metal is lost. If V = volume,

$$V_1 = l_1 \times a_1 = l_1 \times \frac{\pi d_1^2}{4}$$

and
$$V_2 = l_2 \times a_2 = l_2 \times \frac{\pi d_2^2}{4}$$

But
$$V_1 = V_2, \text{ so } l_1 \times \frac{\pi d_1^2}{4} = l_2 \times \frac{\pi d_2^2}{4}$$

Therefore
$$l_2 = l_1 \times \frac{\pi d_1^2}{4} \times \frac{4}{\pi d_2^2} = l_1 \times \frac{d_1^2}{d_2^2}$$

Since the wire is one-fifth of the initial diameter after drawing,

$$\frac{d_1}{d_2} = \frac{1}{1/5} = 5$$

Therefore
$$l_2 = 10 \times \frac{5^2}{1^2} \text{ metres} = 10 \times 25 \text{ metres} = 250 \text{ m}$$

$$R \propto \frac{l}{a}, \text{ so } R_2 = R_1 \times \frac{l_2}{l_1} \times \frac{a_1}{a_2}$$

$$R_1 = 0{\cdot}015 \, \Omega, \quad l_1 = 10 \text{ m}, \quad l_2 = 250 \text{ m}$$

$$a_1 = \frac{\pi d_1^2}{4} \propto \frac{\pi 1^2}{4} \propto \frac{\pi}{4}$$

$$a_2 = \frac{\pi d_2^2}{4} \propto \frac{\pi}{4} \times \left(\frac{1}{5}\right)^2 \propto \frac{\pi}{4 \times 25}$$

Therefore
$$R_2 = 0{\cdot}015 \times \frac{250}{10} \times \frac{\pi}{4} \times \frac{4 \times 25}{\pi} \text{ ohms}$$

$$= 0{\cdot}015 \times 25 \times 25 \text{ ohms}$$

$$= 9{\cdot}375 \, \Omega$$

2.3 RESISTIVITY

The resistance between opposite faces of a cube of the conductor material, given the value r ohms in Section 2.2, is called the **resistivity**, or, sometimes, the **specific resistance**, of the material. For most conductors, this value is very low, and is usually measured in microhms ($\mu\Omega$) for a cube of given side length. For instance, the resistivity of copper is about $0\cdot 0172\,\mu\Omega$ for a 1 metre cube, and is therefore expressed as $0\cdot 0172\,\mu\Omega$m. The symbol for resistivity is the Greek letter ρ (pronounced 'rho'), which can now take the place of r in our previous formula, so that

$$R = \frac{\rho l}{a}$$

where
R = conductor resistance, $\mu\Omega$
ρ = cable resistivity, $\mu\Omega$m *or* $\mu\Omega$mm
l = cable length, m *or* mm
a = cable cross-sectional area, m² *or* mm²

For this formula to be correctly applied, it should be noticed that

(a) the resistance calculated will be in the same units as those in which resistivity is given (usually microhms)
(b) the units of length and cross-sectional area must match the unit included in the given resistivity; for instance, if resistivity is quoted in microhm metres, length must be in metres and cross-sectional area in square metres.

Table 4 gives the resistivities, in three commonly used units, of some metallic conductors found in installation work. These are average values, since a slight variation is found to occur depending on conductor condition.

Table 4 Resistivity of common conductors

Material	Resistivity		
	$\mu\Omega$m	$\mu\Omega$cm	$\mu\Omega$mm
Copper (annealed)	0·0172	1·72	17·2
Copper (hard drawn)	0·0178	1·78	17·8
Aluminium	0·0285	2·85	28·5
Tin	0·114	11·4	114
Lead	0·219	21·9	219
Mercury	0·958	95·8	958
Iron	0·100	10·0	100
Silver	0·0163	1·63	16·3
Brass	0·06–0·09	6–9	60–90

The figures given in Table 4 will usually increase as the material temperature increases. This effect will be considered in Section 2.5.

The differing values of resistivity for different conductor materials make it clear that the resistance of a wire or cable will depend on the material of which it is made, as well as on its length and cross-sectional area. Table 4 thus shows clearly why copper is so widely used as a conductor material, being second only to silver in its resistivity value. Aluminium is about half as resistive again as copper, but its lightness and cheapness have led to its increasing use for cables and busbars.

2.4 RESISTANCE CALCULATIONS

All resistance calculations are based on the formula $R = \rho l/a$, given in Section 2.3. Thus, if resistivity, length and cross-sectional area are known, the resistance may be calculated. In fact, if any three of the four values are known, the fourth may be found.

EXAMPLE 2.5

Calculate the resistance of 1000 m of 15 mm² single copper conductor. (The cross-sectional area of the conductor is 15 mm².)

The three quantities ρ, l and a must be expressed in terms of the same unit of length. If the millimetre is chosen,

$$\rho = 17\cdot 2\,\mu\Omega\text{mm from Table 4, or } \frac{17\cdot 2}{10^6}\,\Omega\text{mm}$$

$$l = 1000\,\text{m} = 1000 \times 1000\,\text{millimetres} = 1\,000\,000\,\text{mm or } 10^6\,\text{mm}$$

$$a = 15\,\text{mm}^2$$

$$R = \frac{\rho l}{a}$$

$$= \frac{17\cdot 2}{10^6} \times \frac{10^6}{15}\,\text{ohms}$$

$$= \frac{17\cdot 2}{15}\,\text{ohms}$$

$$= 1\cdot 15\,\Omega$$

As an alternative, we could choose the metre.

$$\rho = 0\cdot 0172\,\mu\Omega\text{m from Table 4.}$$

$$l = 1000\,\text{m}$$

$$a = 15\,\text{mm}^2 = \frac{15}{10^6}\,\text{square metres}$$

$$R = \frac{\rho l}{a}$$

$$= \frac{0\cdot 0172}{10^6} \times 1000 \times \frac{10^6}{15}\,\text{ohms}$$

$$= \frac{17\cdot 2}{15}\,\text{ohms}$$

$$= 1\cdot 15\,\Omega$$

EXAMPLE 2.6

Calculate the resistance of 1000 m of hard-drawn copper conductor, if ρ for the material is $0\cdot 0178\,\mu\Omega\text{m}$ and the cross-sectional area of the conductor is $4\cdot 7\,\text{mm}^2$.

Since ρ is given in $\mu\Omega\text{m}$, l must be in metres and a in square metres.

$$R = \frac{\rho l}{a} = \frac{0\cdot 0178 \times 1000}{10^6} \times \frac{10^6}{4\cdot 7}\,\text{ohms}$$

$$= \frac{17\cdot 8}{4\cdot 7}\,\text{ohms}$$

$$= 3\cdot 79\,\Omega$$

EXAMPLE 2.7

Calculate in metres the length of an aluminium conductor (of cross-sectional area $40\,\text{mm}^2$) which will have a resistance of $0\cdot 57\,\Omega$.

From Table 4, ρ for aluminium is $0\cdot 0285\,\mu\Omega\text{m}$, so a must be expressed in m^2, when l will be found in m.

$$R = \frac{\rho l}{a} \text{ so } l = \frac{Ra}{\rho}$$

Therefore

$$l = 0\cdot 57 \times \frac{40}{10^6} \times \frac{10^6}{0\cdot 0285}\,\text{metres}$$

$$= \frac{0\cdot 57 \times 40}{0\cdot 0285}\,\text{metres}$$

$$= 800\,\text{m}$$

EXAMPLE 2.8

The heating spiral of an electric fire is removed from its firebar, uncoiled and measured. It has a length of 2 m, and a diameter of 0·2 mm, and its resistance is measured as 10 Ω. What is the resistivity, in μΩmm, of the spiral material?

$$R = \frac{\rho l}{a} \text{ so } \rho = \frac{Ra}{l}$$

Since ρ is required in μΩmm, R must be expressed in microhms, a in square millimetres and l in millimetres.

$$a = \frac{\pi d^2}{4}$$

$$= \frac{3 \cdot 142 \times 0 \cdot 2 \times 0 \cdot 2}{4} \text{ square millimetres}$$

$$= 0 \cdot 03142 \text{ mm}^2$$

$$\rho = \frac{Ra}{l}$$

$$= \frac{10 \times 1\,000\,000 \times 0 \cdot 03142}{2000} \text{ microhm millimetres}$$

$$= \frac{314 \cdot 2}{2} \text{ microhm millimetres}$$

$$= 157 \cdot 1 \text{ μΩmm}$$

The examples above indicate the steps to be followed in this type of problem, which has many practical applications. The golden rule is to be careful to express resistivity, length and cross-sectional area in terms of the same unit of length.

2.5 EFFECT OF TEMPERATURE ON RESISTANCE

The following test will show in a very convincing manner that change of temperature can affect the resistance of a conductor. When connected to a 2 V supply, a 60 W, 240 V lamp takes a current of 25 mA. Its resistance can thus be calculated.

$$R = \frac{V}{I} = \frac{2}{0 \cdot 025} \text{ ohms} = 80 \, \Omega$$

If the lamp is now connected to a 240 V supply, its filament becomes white hot and glows brightly, and will be found to take a current of 250 mA. Its new resistance will thus be

$$R = \frac{V}{I} = \frac{240}{0 \cdot 250} \text{ ohms} = 960 \, \Omega$$

We can see from this that the increase in temperature has increased the filament resistance by 12 times! This extreme example is striking proof that temperature affects resistance.

It is often important to be able to calculate the resistance of a conductor at any given temperature. To be able to achieve this simply, we use the **temperature coefficient of resistance** for the material concerned, which is given the symbol α (Greek 'alpha'). The temperature coefficient of resistance of a material at 0°C can be defined as the change in resistance of a 1 Ω sample of the given material, when its temperature is increased from 0°C to 1°C. For instance, an annealed copper conductor, which has a resistance of 1 Ω at 0°C, will have a resistance of 1·0043 Ω at 1°C. Hence we can say that the temperature coefficient of resistance of annealed copper at 0°C is 0·0043 ohm per ohm per degree centigrade, or 0·0043/°C.

In practice, it is often difficult to measure resistances at 0°C, so temperature coefficients of resistance are often expressed for other temperatures, such as 20°C. Table 5 gives some temperature coefficients of resistance of some common conductors.

Carbon has a negative temperature coefficient of resistance, which means that its resistance decreases as temperature increases.

Table 5 Temperature coefficients of resistance of some conductors

Material	(/°C at 0°C)	(/°C at 20°C)
Annealed copper	+ 0·0043	+ 0·00396
Hard-drawn copper	+ 0·0043	+ 0·00396
Aluminium	+ 0·0040	+ 0·00370
Brass	+ 0·0010	+ 0·00098
Iron	+ 0·0066	+ 0·00583
Nickel-chromium	+ 0·00017	+ 0·000169
Carbon (graphite)	− 0·0005	− 0·00047
Silver	+ 0·0041	+ 0·00379

Provided that the temperature coefficient of resistance of a conductor material is known, a simple formula can be deduced for calculating the conductor resistance at any temperature from the resistance at 0°C. Let R_0 be the resistance of a conductor at 0°C, and let α be its temperature coefficient of resistance in '/°C' at 0°C.

From the definition of temperature coefficient of resistance, the change in resistance will be R_0 ohms at 1°C, $2R_0$ ohms at 2°C and tR_0 ohms at t degree Centigrade. If R_t is the total conductor resistance at t(°C),

$$R_t = R_0 + R_0 \alpha t$$

or

$$R_t = R_0(1 + \alpha t)$$

If the temperature coefficient of resistance at, say, 20°C were being used, the formula could be written as $R_t = R_{20}(1 + \alpha t)$, where R_{20} is the conductor resistance at 20°C, t is the change of temperature from 20°C, and α is the temperature coefficient of resistance at 20°C.

EXAMPLE 2.9

The resistance of 1000 m of 2·5 mm² annealed copper conductor is 5·375 Ω at 20°C. Find its resistance at 50°C.

From Table 5, α is 0·00396/°C at 20°C.

Therefore
$$R_t = R_0(1 + \alpha t)$$
$$= 5·375\{1 + 0·00396 (50-20)\} \text{ ohms}$$
$$= 5·375 \times 1·1188 \text{ ohms} = 6 \, \Omega$$

One method of temperature measurement is to subject a resistor to the unknown temperature and accurately measure its resistance, hence calculating the temperature from a knowledge of the temperature coefficient of resistance of the resistor material.

EXAMPLE 2.10

A resistance thermometer with a temperature coefficient of 0·001/°C at 0°C and a resistance of 3 Ω at 0°C is placed in the exhaust gases of an oil-fired furnace. The thermometer resistance rises, reaching a final steady value of 5·25 Ω. What is the exhaust-gas temperature?

$$R_t = R_0(1 + \alpha t)$$
$$5·25 = 3(1 + 0·001t)$$
$$5·25 = 3 + 0·003t$$
$$5·25 - 3 + 0·003t$$

Therefore
$$t = \frac{2·25}{0·003} °C = 750°C$$

There are many practical cases where the temperature coefficient of resistance of a material is known at a given temperature, but where it is impracticable to measure the actual resistance of some particular conductor at this temperature. If R_1 is the conductor resistance at temperature t_1, and R_2 the resistance at temperature t_2.

$$R_1 = R_0(1 + \alpha t_1) \quad \text{and} \quad R_2 = R_0(1 + \alpha t_2)$$

Dividing these equations, we have

$$\frac{R_1}{R_2} = \frac{R_0(1 + \alpha t_1)}{R_0(1 + \alpha t_2)} = \frac{1 + \alpha t_1}{1 + \alpha t_2}$$

Hence

$$R_2 = \frac{R_1(1 + \alpha t_2)}{1 + \alpha t_1}$$

and R_0, the resistance at $0°C$, is not needed.

EXAMPLE 2.11

The field winding of a d.c. motor is of annealed copper, and has a resistance of 500 Ω at 15°C. What field current will flow at the operating temperature of 35°C if the field p.d. is 300 V?

From Table 5, α for annealed copper is 0·0043/°C at 0°C.

$$R_2 = \frac{R_1(1 + \alpha t_2)}{1 + \alpha t_1}$$

$$= \frac{500(1 + 0\cdot0043 \times 35)}{1 + 0\cdot0043 \times 15} \text{ohms}$$

$$= \frac{500 \times 1\cdot1505}{1\cdot0645} \text{ohms}$$

$$= 540 \,\Omega$$

$$I = \frac{V}{R} = \frac{300}{540} \text{ampere} = 0\cdot556 \text{ A}$$

2.6 EFFECTS OF TEMPERATURE CHANGES

There are some applications, like the resistance thermometer considered in example 2.10, where a resistive conductor is deliberately subjected to temperature changes. Resistors of special materials which give a considerable change in resistance for a small temperature change are often used, and are called **thermistors**. These devices are used increasingly in temperature measuring equipment, and in conjunction with electronic circuits in special types of thermostat.

The change in resistance of a filament lamp described at the beginning of Section 2.5 illustrates clearly how the change in resistance may cause problems. Taking the 60 W, 240 V lamp mentioned, the cold resistance is 80 Ω. At the instant of switching on the 240 V supply, the current to the lamp will be given by

$$I = \frac{V}{R} = \frac{240}{80} \text{amperes} = 3 \text{ A}$$

At this instant, the lamp will be taking about 12 times its normal current! The lamp quickly heats up, increasing in resistance as it does so and reducing the current to its normal value of 250 mA. The short-lived heavy current, called a **transient**, can cause difficulties. The magnetic forces set up within the lamp itself may cause it to fail at the instant of switching, and a weak fuse may blow when a bank of lamps served by it is switched on.

This effect is not often noticeable because of the comparatively small currents taken by filament lamps. It would be serious, however, if the element of an electric fire were wound with tungsten, which is the material used for lamp filaments. The normal running current of a 1 kW fire element connected to a 240 V supply is about 4·2 A, but would be about 40 A at the instant of switching on a tungsten element. A fire element takes some time to reach its final temperature, so the heavy current would pass through the circuit for a period long enough to blow the circuit fuse.

To prevent this state of affairs, the elements of heavily loaded heating equipment, such as fires, cookers and immersion heaters, are wound with a special alloy of nickel and chromium, which has a near-zero temperature coefficient of resistance. This means that there will be almost no change in the resistance of the element as its temperature increases, and hence virtually no change in the circuit current.

There are many circumstances in which changes in the temperature of electrical equipment will affect its operation. For example, the windings of an electric motor will increase in temperature when it is used. This will result in an increase in the resistance of the winding, and a reduction in the current carried, which may affect the operation of the motor. For instance, reduction of current in the field winding of a direct-current motor will result in an increase in its speed.

The resistance of cables will increase with increasing temperature, giving an increased cable-voltage drop (Section 2.7). Usually, however, the cable resistance will form a very small proportion of total circuit resistance, and the effect will be unimportant.

2.7 VOLTAGE DROP IN CABLES

When choosing cables for an installation, it is necessary to ensure that they will carry the load current without overheating. An equally important factor which must be considered is the voltage drop that will occur in them owing to current flowing through their resistance. If this voltage drop is excessive, the potential difference across the load will be low and efficient operation will be prevented. 'The I.E.E. Wiring Regulations', published by the Institution of Electrical Engineers lays down the allowable voltage drop from the supply position to any point in the installation.

Since any voltage can be expressed by multiplying a current by a resistance ($V = IR$), voltage drop in a cable depends on the current it carries and on its resistance, the latter depending on conductor material, cable length, and cross-sectional area as well as its temperature (see Section 2.3). The resistance of a given length of cable, and hence its voltage drop for a given current, can be reduced by replacing it with a cable having a larger cross-sectional area.

EXAMPLE 2.12

A twin $2\,mm^2$ m.i. cable feeds a heater which takes a current of 20 A. If the cable is 100 m long, calculate the voltage drop in it, and the p.d. across the heater if the supply voltage is 240 V. What must be the minimum cross-sectional area (c.s.a.) of a replacement cable if the voltage drop is not to exceed 6 V?

M.I. cables have hard-drawn copper cores, so ρ can be taken from Table 4 as $17{\cdot}8\,\mu\Omega mm$. The cross-sectional area of each conductor is $2\,mm^2$, and the total conductor length will be 200 m.

Cable resistance,
$$R = \frac{\rho l}{a}$$

Therefore
$$R = \frac{17\cdot 8 \times 200 \times 1000}{2} \text{ microhms}$$

$$R = 1\,780\,000\,\mu\Omega \text{ or } R = 1\cdot78\,\Omega$$

Cable voltage drop
$$V = I \times R$$
$$= 20 \times 1\cdot78 \text{ volts}$$
$$= 35\cdot6\,V$$

p.d. across heater
$$= 240 - 35\cdot6 \text{ volts} = 204\cdot4\,V$$

Clearly, the heater, designed for operation at 240 V, would be very inefficient with such a low applied voltage. For this reason, the IEE Wiring Regulations limit the maximum allowable voltage drop in a circuit to $2\frac{1}{2}\%$ of the supply voltage, or 6 V for a 240 V supply.

A voltage drop of $35\cdot6\,V$ is far in excess of the allowable limit of 6 V, even though the cable current-carrying capacity of 26 A is not exceeded. If the voltage drop must not exceed 6 V, the maximum cable resistance can be found thus:

$$R = \frac{V}{I} = \frac{6}{20} \text{ ohms} = 0\cdot3\,\Omega$$

$$R = \frac{\rho l}{a}, \text{ so } a = \frac{\rho l}{R}$$

Minimum c.s.a.
$$a = \frac{17\cdot 8 \times 200 \times 1000}{1\,000\,000 \times 0\cdot3} \text{ square millimetres}$$

$$= 11\cdot9\,mm^2$$

This is the minimum c.s.a. to satisfy the voltage-drop requirements of the Regulations, but it is not a standard size. The next standard size above this is 15 mm² twin m.i. cable (from the IEE Wiring Regulations tables). The fact that the current rating of this cable is 90 A, whereas it is only required to carry 20 A, illustrates the importance of considering voltage drop as well as current-carrying capacity when choosing cables.

Tables of cable current ratings in the IEE Wiring Regulations include figures for the voltage drop per ampere per metre length of run for the cable concerned under the conditions indicated. The IEE tables are more accurate than the method in the above example, since, for a.c. circuits, they take into account cable reactance as well as resistance. Reactance and its combination with resistance will be considered in Chapter 10.

In some circumstances, it may be necessary to calculate the resistance of a supply cable to find the voltage drop in it. This process is illustrated in the following examples:

EXAMPLE 2.13

A motor takes 45 A from a 240 V supply. The motor is fed from the supply by a twin aluminium cable 40 m long, each core of the cable having a cross-sectional area of 20 mm². Calculate the voltage at the motor terminals.

The first step is to find the cable resistance. The total length of conductor ('go' and 'return') is $2 \times 40 \text{ m} = 80 \text{ m}$, or 80 000 mm, and, from Table 4, the resistivity of aluminium is 28·5 $\mu\Omega$mm.

$$R = \frac{\rho l}{a}$$

$$= \frac{28 \cdot 5 \times 80\,000}{1\,000\,000 \times 20} \text{ ohms}$$

$$= 0 \cdot 114 \, \Omega$$

$$\text{Voltage drop} = I \times R$$

$$= 45 \times 0 \cdot 114 \text{ volts}$$

$$= 5 \cdot 1 \text{ V approximately}$$

$$\text{Motor voltage} = \text{supply voltage} - \text{voltage drop}$$

$$= 240 - 5 \cdot 1 \text{ volts}$$

$$= 234 \cdot 9 \text{ V}$$

EXAMPLE 2.14

An industrial heater is fed from the supply by a twin annealed-copper cable with 2·5 mm² conductors. The current to the heater is 20 A, and the supply voltage is 240 V. If the potential difference at the heater terminals is 235·5 V, calculate the length of the supply cable to the nearest metre.

Here, we must first find the cable voltage drop, and thus its resistance.

$$\text{Voltage drop} = 240 - 235 \cdot 5 \text{ volts}$$

$$= 4 \cdot 5 \text{ V}$$

$$\text{Cable resistance} = \frac{\text{voltage drop}}{\text{current}}$$

$$= \frac{4 \cdot 5}{20} \text{ ohms}$$

$$= 0 \cdot 225 \, \Omega$$

$$R = \frac{\rho l}{a}, \text{ so changing the subject of the formula,}$$

$$l = \frac{Ra}{\rho}$$

Therefore
$$l = \frac{0.225 \times 2.5 \times 10^6}{17.2} \text{ millimetres}$$

(from Table 4, ρ for annealed copper is $17.2\,\mu\Omega\text{mm}$)

$$l = \frac{0.225 \times 2.5}{17.2} \times 10^3 \text{ metres}$$

$$= 32.7\,\text{m}$$

This is the total conductor length for the twin cable. The cable length is half the conductor length, which, to the nearest metre, is 16 m.

2.8 SUMMARY OF FORMULAS FOR CHAPTER 2

$$R = \frac{\rho l}{a} \qquad a = \frac{\rho l}{R} \qquad \rho = \frac{Ra}{l} \qquad l = \frac{Ra}{\rho}$$

where
- R = conductor resistance, $\mu\Omega$
- ρ = conductor resistivity, $\mu\Omega\text{m}$ or $\mu\Omega\text{mm}$
- l = conductor length, m or mm
- a = conductor cross-sectional area, m^2 or mm^2

Note that, if, for example, ρ is used in $\mu\Omega\text{m}$, l must be expressed in m and a in m^2.

$$R_t = R_0(1 + \alpha t) \qquad R_0 = \frac{R_t}{1 + \alpha t}$$

$$\alpha = \frac{R_t - R_0}{R_0 t} \qquad t = \frac{R_t - R_0}{R_0 \alpha}$$

where
- R_t = resistance at t degrees Celsius
- R_0 = resistance at 0°C
- α = temperature coefficient of resistance, per deg C at 0°C

$$R_2 = \frac{R_1(1 + \alpha t_2)}{1 + \alpha t_1}$$

where
- R_1 = resistance at t_1 degrees Celsius
- R_2 = resistance at t_2 degrees Celsius

Cable voltage drop $\quad V = IR$

where
- R = total cable resistance, Ω
- I = cable current, A

$$V_L = V_S - V$$

where
- V_L = load terminal voltage
- V_S = supply voltage
- V = cable voltage drop

2.9 EXERCISES

1. The resistance of 100 m of $4.5\,\text{mm}^2$ cable is $0.36\,\Omega$. What is the resistance of 600 m of this cable?

2. A single-core cable, 24 m long, has a measured conductor resistance of $0.06\,\Omega$. What is the resistance of 1000 m of this cable?

3. The resistance of $15\,\text{mm}^2$ single cable is $1.21\,\Omega/\text{km}$. Calculate the resistance of a twin $15\,\text{mm}^2$ cable, 80 m long, if the cores are connected in series.

4. 100 m of $40\,\text{mm}^2$ copper cable has a resistance of $0.0445\,\Omega$. What is the resistance of 100 m of $15\,\text{mm}^2$ copper cable?

5. 1 km of $4.5\,\text{mm}^2$ copper cable has a resistance of $3.82\,\Omega$. Calculate the resistance of 250 m of $2\,\text{mm}^2$ copper cable.

6. If the resistance of $2\,\text{mm}^2$ copper cable is $8\,\Omega/\text{km}$, what length of $1\,\text{mm}^2$ copper cable will have a resistance of $4\,\Omega$?

7. Copper cable of cross-sectional area $40\,\text{mm}^2$ has a resistance of $0.445\,\Omega/\text{km}$. What will be the cross-sectional area of a similar cable, 100 m long, which has a resistance of $0.89\,\Omega$?

8   A certain cable has a resistance of 0·5 Ω. What is the resistance of a cable made of the same material which has twice the cross-sectional area and is three times as long as the first?

9   Two cables have equal resistances, but one has a cross-sectional area 2·5 times greater than the other. How much longer is the thicker cable than the thinner cable?

10 In the manufacturer of copper wire, a thick circular rod, which has a resistance of 0·01 Ω, is drawn out without change in volume until its diameter is one tenth of what it was. What is its new resistance? (C & G)

11 How do the dimensions of an electrical conductor affect its resistance? (NCTEC)

12 (a) How does the electrical resistance of a copper conductor vary with
            (i) an increase in its length
            (ii) an increase in its cross-sectional area
            (iii) an increase in its temperature?
    (b) Place the following materials in ascending order of resistivity.
         glass, copper, iron, carbon, brass. (C & G)

13 (a) State how the length and the cross-sectional area of a conductor affects its resistance.
    (b) A piece of silver wire and a piece of resistance wire have identical dimensions. State how their electrical resistance will differ, giving the reason for your answer. (ULCI)

14 A twin cable with aluminium conductors with cross-sectional area of 25 mm^2 is 150 m long. What is the resistance of the cable? The resistivity of aluminium is 28·5 $\mu\Omega$mm.

15 Calculate the resistance of 1000 m of copper conductor of 4·5 mm^2 cross-sectional area. Take the resistivity of copper as 17·2 $\mu\Omega$mm.

16 What is the resistance of 200 m of copper conductor with a cross-sectional area of 2 mm^2? The resistivity of copper is 17·2 $\mu\Omega$mm.

17 A cable has a cross-sectional area of 1 mm^2. What length of this cable (to the nearest metre) will have a resistance of 2 Ω? The cable is made of copper with a resistivity of 17·2 $\mu\Omega$mm.

18 An electric-fire element has a total length of 10 m, a cross-sectional area of 0·5 mm^2 and a resistance of 50 Ω. Calculate the resistivity of the material from which it is made.

19 What is the cross-sectional area of a cable having a resistivity of 20 $\mu\Omega$mm if its resistance is 1·5 Ω/km?

20 Calculate the length of an aluminium cable of cross-sectional area 125 mm^2 which has a resistance of 0·1 Ω. The resistivity of aluminium is 28·5 $\mu\Omega$mm.

21 500 m of cable is buried and cannot be examined. The cross-sectional area of the cable is known to be 3 mm^2, and its resistance is measured as 2·8 Ω. Suggest the material from which the conductor is made.

22 The lead sheath of a cable, 600 m long, has a measured resistance of 20 Ω. If the resistivity of lead is 219 $\mu\Omega$mm, find the cross-sectional area of the sheath.

23 The shunt-field winding of a d.c. motor has a resistance of 100 Ω at 20°C. What will be its resistance at 45°C, if it is made of copper? (α = 0·00396/°C at 20°C)

24 The resistance of a steel catenary wire is 2 Ω at 20°C, but is used to support lighting fittings in a large coldroom which is kept at a temperature of −10°C. What will be its cold resistance, if α for steel is 0·006/°C at 20°C?

25 The resistance of 1 mm^2 copper conductor is 13 Ω/km at 20°C. What will be its resistance at 60°C? (α = 0·00396/°C at 20°C)

26 A 1 mm^2 m.i. cable has a resistance of 1·5 Ω at 20°C. When installed to feed control circuits on a furnace, its resistance is measured as 2·5 Ω. What is the average operating temperature? α for copper is 0·00396/°C at 20°C.

27 A resistance thermometer has a temperature coefficient of 0·002/°C at 0°C and a resistance of 40 Ω. When used to measure the temperature of an oven, the resistance increases to 56 Ω. What is the temperature of the oven?

28 The resistance of a coil of cable at 10°C is 3·5 Ω. What will be its resistance at 25°C? The cable conductor is made of aluminium, for which α is 0·0040/°C at 0°C.

29 If a resistance thermometer is made of brass wire and has a resistance of 10 Ω at 15°C, what will be its resistance at 300°C? α for brass is 0·0010/°C at 0°C.

30 A motor is used to drive a conveyor feeding a stove-enamelling plant, and its temperature when idle never falls below 30°C. At this temperature, the winding resistance is measured as 15 Ω. When running, the operating temperature of the motor is 65°C. What is the winding resistance at the higher temperature? The winding is of copper with α = 0·0043/°C at 0°C.

31 A voltmeter coil has a resistance of 1000 Ω. The meter indicates correctly at 15°C. If the coil temperature rises to 65°C and the temperature coefficient of the coil wire is 0·0004/°C, what reading will it show with 10 V applied?

32 The copper shunt-field coil of a d.c. motor takes 2·5 A from a 200 V supply when it is first switched on and its temperature is 16°C. What will be the field current when the motor has been running for some time and its temperature increases to 56°C? Take α for copper as 0·0043/°C at 0°C.

33 Calculate the resistance of 200 m of single-core 1 mm^2 m.i. cable, with hard-drawn copper conductor, at 20°C. What will be the resistance of this cable at 80°C? ρ for hard-drawn copper at 20°C is 17·8 μΩmm. α for hard-drawn copper is 0·00396/°C at 20°C.

34 A resistor used in a measuring circuit has a resistance of 10 kΩ at 20°C, and is made of a material having a temperature coefficient of resistance of 0·0001/°C at 20°C. What is the percentage change in the value of the resistor when its temperature is 30°C?

35 The maximum permissible voltage drop in an installation is 2·5% of the supply voltage. What will be the minimum load voltage, to the nearest volt, for supply voltages of
   (a) 200 V,
   (b) 230 V,
   (c) 240 V, and
   (d) 250 V?

36 Each core of a twin cable carrying a current of 10 A has a resistance of 0·15 Ω. What is the total voltage drop in the cable?

37 An industrial heater is connected to a 240 V supply by means of a twin cable, each core having a resistance of 0·08 Ω. The heater takes a current of 25 A from the supply. What is the p.d. across its terminals?

38 What must be the maximum resistance of each core of a twin cable, feeding a heater taking 12 A from a 240 V supply, if the allowable voltage drop of 2·5% of supply voltage must not be exceeded?

39 A laboratory socket outlet is fed through a twin cable with each core having a resistance of 0·2 Ω. The supply is at 115 V, and the cable voltage drop must not exceed 1·5% of this value. What maximum current should be taken from the outlet?

40 A motor is fed from 240 V mains 50 m away by a 2-core cable, each core of which has a cross-sectional area of 8·6 mm^2. Calculate the voltage at the motor terminals when it is taking 30 A. Take the resistivity of copper as 17·2 μΩmm.

41 An industrial oil heater takes a current of 60 A from a 400 V supply through a twin-core aluminium cable, 80 m long. If the cable voltage drop must not exceed 2½% of the supply voltage, calculate the minimum cross-sectional area of each core. Take ρ for aluminium as 28 μΩmm.

42 An immersion heater takes a current of 12·5 A and is fed through a twin cable, each core of which has a cross-sectional area of 2·5 mm^2. The cable conductors are made of copper, which has a resistivity of 17·2 μΩmm. Calculate the greatest length of cable which may be used if the cable voltage drop must not exceed 6 V.

43 A motor takes 10 A from a 200 V supply, and is fed through a twin 2·5 mm^2 copper cable 30 m long. Calculate the voltage at the motor terminals. ρ for copper is 17·2 μΩmm.

Chapter 3

Mechanics

At first sight, it may seem strange that a book dealing with electrical theory should concern itself with mechanics. What connection is there?

As well as understanding the operation of electrical equipment, an electrical craftsman has to manufacture and install it. This involves various mechanical operations, such as cutting cables, threading conduits, driving screws, lifting heavy apparatus, and so on. An understanding of the principles involved can hardly fail to reduce both the physical effort required, and the likelihood of accident, in carrying out these tasks.

3.1 MASS, FORCE, PRESSURE AND TORQUE

Mass
Mass can be defined as the amount of material in an object, and is usually measured by comparison with another mass chosen as a standard. For instance, the standard of mass in SI units is the **kilogram** (kg), which is the mass of a block of platinum kept at Sèvres in France.

Force
A scientific definition of force is difficult at this stage, but it will suffice to say that force can be measured in terms of the effects it produces. A force can lift, bend or break an object. It can move an object previously stationary, or can stop or change the rate and direction of movement. In electrical-craft practice, force is used to move, bend, cut or join together materials, as well as to drive in screws and perform a host of other tasks.

Weight
The Earth exerts a natural force of attraction on all other masses. This force is generally called the 'gravitational pull' of the Earth on a body, and the magnitude of this force measures the weight of the body.

Units of force and weight
A mass of one kilogram will experience a force due to the Earth's gravity of 1 kilogram force. However, since the kilogram is a unit of mass, it should not be used to measure force. The SI unit of force is the **newton** (symbol N). A mass of one kilogram experiences a force due to gravity of 9·81 newtons.

We could say, for example, that a piece of machinery has a mass of 1 kg, in which case it will require a force of 9·81 N to lift it against gravity. If the machinery formed part of a space rocket, it could be sent outside the gravitational pull of the Earth. Its mass would then be unchanged, but it would experience no earth gravitational force.

EXAMPLE 3.1

A bundle of conduit has a mass of 800 kg. What force will be needed to lift it?

1 mass of 1 kg requires a force of 9·81 N to lift it against gravity.

$$\text{Force required} = 800 \times 9\cdot81 \text{ newtons}$$
$$= 7848 \text{ N}$$

Pressure
The pressure on a surface is measured in terms of **the force per unit area on the surface,** assuming that the surface is at right angles to the direction of the force. For example, a force of 500 N acting on one leg of a tripod pipevice with a sharp edge of effective area 20 mm² in contact with a floor will exert a pressure given by

$$\text{pressure} = \frac{\text{force}}{\text{effective area}} = \frac{500 \text{ N}}{20 \text{ mm}^2} = 25 \text{ N/mm}^2$$

A pressure as high as this will damage many floor finishes. If the leg of the vice is stood on a piece of wood 100 mm square, the same force is spread over an area of 10 000 mm², and the pressure falls to

$$\frac{500 \text{ N}}{10\,000 \text{ mm}^2} = 0\cdot05 \text{ N/mm}^2$$

Strictly speaking, the SI unit of pressure is the newton per square metre or pascal. Since there are 1 000 000 mm² in 1 m², the pressure of 0·05 N/mm² given above could also be expressed as 5×10^4 N/m².

Density

If we take two blocks of the same size, one made of wood and the other of iron, the iron block will be heavier than the wooden block. Thus we can say that iron is more dense than wood. The density of a material is the mass of it which is contained in one cubic unit, and is given in kilograms per cubic metre (kg/m³).

EXAMPLE 3.2

What will be the mass of a block of copper 0·1 m by 0·2 m by 0·15 m? The density of copper is 8900 kg/m³.

The volume of the block is

$$0.1 \times 0.2 \times 0.15 = 0.003 \text{ m}^3$$

$$\text{Since density} = \frac{\text{mass}}{\text{volume}}$$

$$\text{mass} = \text{density} \times \text{volume}$$

$$= 8900 \times 0.003 \text{ kilograms}$$

$$= 26.7 \text{ kg}$$

Torque

In some cases, a force may tend to cause rotary movement. Probably the simplest example of this principle in electrical-craft work is the process of threading a conduit with a stock and die. The turning effect is called the **torque**, or, sometimes, the **turning moment of a force**, and is measured as the force applied multiplied by the perpendicular distance between the direction of the force and the point about which rotation can occur. Fig. 3.1 makes it clear that, to obtain maximum torque for a given applied force, the force should always be in the direction of the resulting movement.

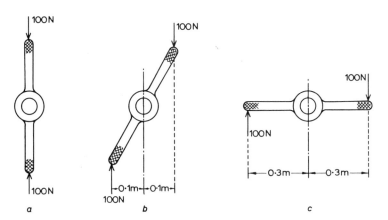

Fig. 3.1 Torque applied to stocks and dies
    a Torque = 100 N × 0 × 2 = zero
    b Torque = 100 N × 0·1 m × 2 = 20 Nm
    c Torque = 100 N × 0·3 m × 2 = 60 Nm

Torque (or turning moment) is measured in newton metres (Nm). In Fig. 3.1a, there is no torque; although there is a force of 100 N, this is trying to compress the stock and not to rotate it, since the distance between the direction of the force and the turning point is zero. The torque produced in Fig. 3.1b is less than that in Fig. 3.1c. The forces are the same in both cases, but, in the latter example, the distance between the direction of the force and the turning point is maximum. In other words, all the force applied is being used to turn the stock, and none is wasted in trying to distort the stock. The actual torque required to thread a conduit depends on its size, but can be reduced considerably by the use of a sharp die and correct lubrication. For a given torque, the force required can be reduced if exerted at a greater radius from the turning point (example 3.3).

44 Mechanics

EXAMPLE 3.3

A conduit requires a torque of 20 Nm for the threading operation. What minimum force must be exerted by each hand pressing at the ends of a stock of overall length (*a*) 0·5 m, (*b*) 1 m?

Each hand provides half the total torque, so the torque per hand will be 10 Nm.

$$\text{Since torque} = \text{force} \times \text{direct distance}$$

$$\text{force} = \frac{\text{torque}}{\text{direct distance}}$$

In both cases, the direct distance from the point of application of the force to the turning point will be half the overall length of the stock.

(*a*) Minimum force required $= \dfrac{10\,\text{Nm}}{0\cdot25\,\text{m}} = 40\,\text{N}$

(*b*) Minimum force required $= \dfrac{10\,\text{Nm}}{0\cdot5\,\text{m}} = 20\,\text{N}$

The lesson drawn from this example should not tempt the craftsman to increase the effective length of his tool handles by the use of a suitable length of pipe or other means. The excessive torque that can then be applied is often sufficient to damage tools and work. A tool will usually be made with handles of such length as to prevent the application of too much torque by a person of normal strength.

3.2 WORK, ENERGY AND POWER

Work

If a force is applied to a body and movement results, work is done. For instance, work is done when a weight is lifted or when forces applied to a stock cause it to rotate.

Work is measured in terms of the distance moved by the object and the force which caused the movement. When the movement is in the same direction as the force, the work done is equal to the distance moved multiplied by the force exerted.

$$\text{Work} = \text{distance} \times \text{force}$$

The unit is the **metre-newton** (mN), which is also known as the **joule (J)**.

EXAMPLE 3.4

A bundle of conduit has a mass of 200 kg. What is its weight? What work must be done in lifting the conduit from the floor on to a rack 2 m high?

Weight is the force exerted on the mass owing to gravity. Since a mass of 1 kg exerts a force due to gravity of 9·81 N,

$$\begin{aligned}
\text{Weight} &= \text{mass} \times 9\cdot81\,\text{newtons} \\
&= 200 \times 9\cdot81\,\text{newtons} \\
&= 1962\,\text{N}
\end{aligned}$$

$$\begin{aligned}
\text{Work done} &= \text{distance} \times \text{force} \\
&= 2\,\text{m} \times 1962\,\text{N} \\
&= 3924\,\text{mN or } 3924\,\text{J}
\end{aligned}$$

EXAMPLE 3.5

A force of 100 N will just move a van on a level road. What work will be expended in pushing the van 15 m?

$$\begin{aligned}
\text{Work} &= \text{distance} \times \text{force} \\
&= 15\,\text{m} \times 100\,\text{N} \\
&= 1500\,\text{mN or } 1500\,\text{J}
\end{aligned}$$

Since the van is not being lifted, its weight is only important in its effect on the force needed to move it against friction.

Energy

Energy is the capacity to do work. It may take many forms, such as nuclear, chemical, heat, mechanical or electrical energy. If we ignore the theoretical atomic physics, which never affects electrical-craft work, it is true to say that, whereas energy can be converted from one form to another, it can be neither created nor destroyed. For instance, coal or oil, containing chemical energy, is burned in the boilers of a power station, and produces heat. This heat evaporates water to become steam under pressure, which is fed to a turbine where mechanical energy is produced. The turbine drives an alternator which produces electrical energy.

The units of energy are the same as those of the work it is capable of performing, i.e. the metre newton (joule).

Efficiency

Although energy cannot be created or destroyed, it does not follow that energy can be converted from one form to another without waste. For instance, the conversions in a power station generally result in little more than one-third of the available chemical energy becoming electrical energy. The difference is not destroyed, but is dissipated, largely in the form of heat, and is lost from the process. Improvements in design have reduced these losses considerably in recent years. Thus we can say that power stations have become more efficient, because, for a given input of energy, the output is greater.

Efficiency of mechanical systems can be defined as the ratio of output to input energy or power, and is usually expressed as a percentage, so that

$$\text{Efficiency} = \frac{\text{output energy or power}}{\text{input energy or power}} \times 100\%$$

In most systems, it is more usual to express efficiency in terms of power than energy.

The difference between input and output is the wasted energy, or losses, so that

$$\text{output} = \text{input} - \text{losses}$$

$$\text{input} = \text{output} + \text{losses}$$

Hence two further ways of defining efficiency follow:

$$\text{Efficiency} = \frac{\text{output}}{\text{output} + \text{losses}} \times 100 \text{ percent}$$

$$\text{Efficiency} = \frac{\text{input} - \text{losses}}{\text{input}} \times 100 \text{ percent}$$

Power

Power is defined as the rate of doing work. An electrician using a hammer and chisel can pierce a hole in a brick wall. The same hole could be produced more quickly with a pneumatic drill. In either method, the completed hole will represent the same amount of work, but the rate of doing work is greater in the second method because the pneumatic drill is more **powerful** than the electrician. Thus

$$\text{average power} = \frac{\text{amount of work done}}{\text{time taken to do it}}$$

The unit is the **joule per second** (J/s), which is also known as the **watt (W)**. The watt is too small for many practical applications, so the kilowatt (kW) is often used.

$$1 \text{kW} = 1000 \text{W} = 1000 \text{J/s}$$

EXAMPLE 3.6

An electric motor drives a pump which lifts 1000 l (litres) of water each minute to a tank 20 m above normal water level. What power must the motor provide if the pump is 50% efficient? 1 l of water weighs 9·81 N.

$$\text{Rate of lifting water} = 1000 \times 9 \cdot 81 = 9810 \text{ N/min}$$

Power required by pump = 9810 × 20 = 196 200 mN/min

$$= \frac{196\,200}{60} \text{ newton metres per second (watts)}$$

$$= 3270\,\text{W}$$

Output power of motor $= 3270 \times \dfrac{100}{50}\,\text{W}$

$$= 6540\,\text{W or } 6.54\,\text{kW}$$

3.3 LIFTING MACHINES

A lifting machine is a device which can enable a small effort to lift a large load, and such machines are often necessary in electrical work to move heavy equipment. A few of the simpler types are considered here.

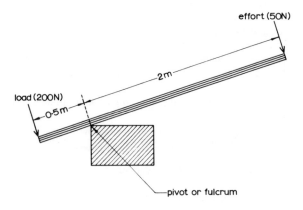

Fig. 3.2 Simple lever

The lever

Consider the simple lever (Fig. 3.2). The total turning moment or torque applied to one end of the lever must be equal and opposite to the torque available at the other end. Thus, if a 50 N effort is applied to the right-hand end of the lever as shown, the total applied torque will be 50 N × 2 m, or 100 Nm. If there are no torque losses (and losses here will be very small) the same torque applied to the right-hand end of the lever will be available at the left-hand end. Since the radius of action about the pivot, or fulcrum, is 0·5 m at this end,

$$100\,\text{Nm} = 0.5\,\text{m} \times \text{available force}$$

$$\text{available force} = \frac{100\,\text{Nm}}{0.5\,\text{m}} = 200\,\text{N}$$

This shows that this lever enables a load of 200 N to be balanced by a force of 50 N, because the lever is four times longer on one side of the fulcrum than it is on the other.

Levers are classified according to the relative positions of the load, the effort and the fulcrum, as shown in Fig. 3.3.

Mechanical advantage and velocity ratio

The lever of Fig. 3.2 has the advantage of allowing a large mass to be lifted by a small force, and mechanical advantage is the ratio of load to effort.

$$\text{Mechanical advantage (m.a.)} = \frac{\text{load}}{\text{effort}}$$

For the lever of Fig. 3.2, for example,

$$\text{m.a.} = \frac{\text{load}}{\text{effort}} = \frac{200\,\text{N}}{50\,\text{N}} = 4$$

This is a ratio, so there is no unit.

For a simple lever, mechanical advantage will be almost constant and, for a class 1 lever, will depend on the lengths of the lever on either side of the fulcrum. For most machines, however, frictional waste will vary with load, and mechanical advantage will not be constant.

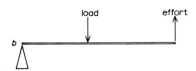

Fig. 3.3 Classification of simple levers
    a Class 1 lever
    b Class 2 lever
    c Class 3 lever

Velocity ratio is defined as the ratio of the distances moved by the effort and the load:

$$\text{velocity ratio (v.r.)} = \frac{\text{the distance moved by the effort in a given time}}{\text{the corresponding distance moved by the load}}$$

For example, it will be clear from Fig. 3.2, that, if the load moved by 1 mm, the effort must move by 4 mm, so that the velocity ratio will be 4 (no units apply).

In an ideal machine (that is, a machine which is 100% efficient),

$$\text{work input} = \text{work output}$$

so

$$\text{effort} \times \text{distance moved by effort} = \text{load} \times \text{distance moved by load}$$

$$\frac{\text{distance moved by effort}}{\text{distance moved by load}} = \frac{\text{load}}{\text{effort}}$$

$$\text{velocity ratio} = \text{mechanical advantage}$$

For most machines, the inefficiency results in a smaller value for mechanical advantage than for velocity ratio, and it is interesting to note that

$$\text{efficiency} = \frac{\text{work out}}{\text{work in}}$$

$$= \frac{\text{load} \times \text{distance moved by load}}{\text{effort} \times \text{distance moved by effort}}$$

$$= \frac{\text{load}}{\text{effort}} \div \frac{\text{distance moved by effort}}{\text{distance moved by load}}$$

$$= \frac{\text{mechanical advantage}}{\text{velocity ratio}}$$

EXAMPLE 3.7

A class 1 lever is arranged so that a load of 1000 N is to be lifted at a distance of 200 mm from the fulcrum. What force must be applied 2 m from the fulcrum to balance the load? Calculate the mechanical advantage and velocity ratio of the system, assuming no losses.

Let F be the required force.

$$\text{Turning moment required for load} = \frac{200}{1000} \times 1000 \text{ newton metres}$$

$$= 200 \, \text{Nm}$$

If there is no waste, this will equal the turning moment to be applied.

Therefore
$$F \times 2 = 200 \, \text{Nm}$$

and
$$F = \frac{200 \, \text{Nm}}{2 \, \text{m}} = 100 \, \text{N}$$

$$\text{m.a.} = \frac{\text{load}}{\text{effort}} = \frac{1000 \, \text{N}}{100 \, \text{N}} = 10$$

$$\text{v.r.} = \frac{\text{effort distance}}{\text{load distance}} = \frac{2 \, \text{m}}{0 \cdot 2 \, \text{m}} = 10$$

Inclined planes and jacks

A simple method of lifting a heavy load is to push it up an inclined plane. For example, heavy machinery can be manhandled on to a lorry with comparative ease if a number of planks form an inclined plane from the ground to the lorry. If necessary, a packing case or trestle can be placed beneath the centre of the planks to give extra support. The load is pushed up the sloping planks in a series of stages. This will involve more work than a direct lift, owing to the friction between the load and the planks, but the force required at any instant is less and rests may be taken between efforts. A useful safety measure to ensure that the load does not slip can be taken by passing a rope, secured at one end to the load, round a solidly fixed structure such as a girder or the lorry chassis. Slack rope is taken up as the load is lifted.

In some machines, a steel inclined plane is, in effect, wound round a central column to form a screw thread. Rotation of the thread can lift a load bearing on it by forcing it up the inclined plane. A machine of this sort is called a jack. One type of jack is shown in Fig. 3.4.

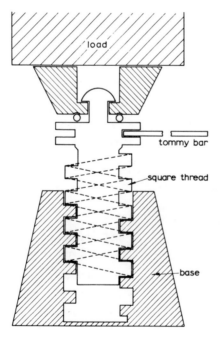

Fig. 3.4 Simple jack

Pulley systems

A machine which is often used to lift heavy loads where a suitable overhead suspension is available is the 'block and tackle'. This consists of two sets of pulleys and a length of rope. A simple arrangement with two blocks each having two pulleys is shown in Fig. 3.5. The pulleys in each block are usually the same size and are mounted side by side, but are drawn as shown for clarity.

Neglecting the rope to which the effort is applied, there are four ropes supporting the load. If the load is to be lifted by, say, one metre, each of these four ropes must shorten by one metre, so that the effort rope must be pulled through four metres. The velocity ratio is thus equal to four in this case, and, in fact, the velocity ratio of a block and tackle is equal to the number of pulleys used. The mechanical advantage (m.a.) would be equal to the velocity ratio (v.r.) if there were no losses, but friction at the pulley bearings reduces its value.

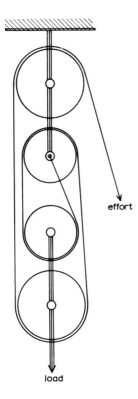

Fig. 3.5 Principle of block and tackle

EXAMPLE 3.8

A block and tackle of two sets, each of five pulleys, is 60% efficient. What is the maximum load which a man weighing 600 N could support using it?

$$\text{v.r.} = 2 \times 5 = 10$$

$$\text{m.a.} = \text{v.r.} \times \text{efficiency} = 10 \times \frac{60}{100} = 6$$

Maximum effort will be applied when the man hangs on the rope. This is an effort of 600 N.

$$\text{Since m.a.} = \frac{\text{load}}{\text{effort}}$$

$$\text{Load} = \text{m.a.} \times \text{effort} = 6 \times 600 \text{ newtons} = 3600 \text{ N}$$

3.4 POWER TRANSMISSION

Pulleys and belts

A method of connecting a motor to a machine is to fit both motor and machine with suitable pulleys, the two connected by a belt. The majority of belts are of the 'vee' type, which grip the pulley and reduce slip.

If the two pulleys are the same size, speeds and torques will be identical if belt slip is neglected. If the pulleys are of differing sizes, a speed change can be arranged. Consider the arrangement shown in Fig. 3.6 where a motor of speed N_1 revolutions per minute is fitted with a pulley of diameter D_1 metres. The motor pulley is coupled by a belt to a machine pulley of diameter D_2 metres and having a speed of N_2 revolutions per minute.

The belt makes contact with both pulleys; so, if we neglect belt slip, the edge speed, or peripheral speed, of each pulley is the same.

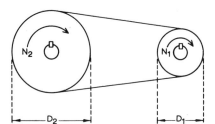

Fig. 3.6 Belt and pulleys. Two pulleys coupled by belt

The motor pulley has a circumference of D_1 metres, and, since it makes N_1 revolutions each minute, its peripheral speed is $\pi D_1 N_1$ metres per minute. Similarly, the peripheral speed of the machine pulley is $\pi D_2 N_2$ metres per minute. Since the peripheral speeds of both pulleys are equal if belt slip is neglected,

$$\pi D_1 N_1 = \pi D_2 N_2$$

so that

$$\frac{N_1}{N_2} = \frac{D_2}{D_1}$$

Thus pulley speeds are in inverse ratio to their diameters.

As well as changing the speed, pulleys of differing sizes also change the torque. It would be wrong to state that no losses occurred in the belt-drive system; but, if we neglect these losses, the work done by the motor is provided to the machine, so that the two are equal. Let the force exerted at the edge of the motor pulley be F_1 newtons, and that the machine pulley be F_2 newtons. In one revolution, the force on the motor pulley moves through πD_1 metres, so that the work done in 1 min is $\pi D_1 N_1 F_1$ joules. Similarly, the work done on the machine pulley in 1 min is $\pi D_2 N_2 F_2$ joules.

Assuming no losses, these values are the same, so that

$$\pi D_2 N_2 F_2 = \pi D_1 N_1 F_1$$

or

$$D_2 F_2 N_2 = D_1 F_1 N_1$$

The torque on the motor pulley is $F_1(D_1/2)$, and that on the machine pulley is $F_2(D_2/2)$.

If motor torque is T_1, and machine torque T_2,

$$T_2 N_2 = T_1 N_1$$

or

$$\frac{T_1}{T_2} = \frac{N_2}{N_1} = \frac{D_1}{D_2}$$

Thus pulley torques are in inverse ratio to their speeds, but in direct ratio to their diameters. If a belt-and-pulley system is used to reduce speed, it must increase torque. Similarly, a pulley-and-belt system intended to increase speed will reduce available torque.

EXAMPLE 3.9

A motor providing a torque of 300 Nm at a speed of 1440 rev/min is fitted with a pulley of diameter 100 mm. This pulley is coupled by a belt to a machine pulley of diameter 400 mm. Assuming no losses or belt slip, calculate the speed of the machine pulley and the torque it can provide.

$$\frac{N_1}{N_2} = \frac{D_2}{D_1}$$

Therefore

$$N_2 = \frac{N_1 D_1}{D_2}$$

Mechanics 51

$$= 1440 \times \frac{100}{400} \text{ revolutions per minute}$$

$$= 360 \text{ rev/min}$$

Also, $$\frac{T_1}{T_2} = \frac{D_1}{D_2}$$

Therefore $$T_2 = \frac{T_1 D_2}{D_1}$$

$$= 300 \times \frac{400}{100} \text{ newton metres}$$

$$= 1200 \text{ Nm}$$

Gearing

Instead of using a belt to connect two pulley wheels, each may have teeth cut in its edge. If the teeth are enmeshed, they must turn together and have the same peripheral speed. Toothed wheels of this sort are called **gears**. They are more positive than a belt as they cannot slip, but may be noisy, and will usually require periodic lubrication.

Since peripheral speeds are the same, the same reasoning applies to gears as to pulleys. It should, however, be noted that two pulleys coupled by a belt rotate in the same direction, whereas two meshed gears rotate in opposite directions. Gear sizes are usually indicated by the number of teeth rather than by diameter, the two being proportional. Thus, if G_1 and G_2 represent the numbers of teeth on two gearwheels respectively,

$$\frac{T_1}{T_2} = \frac{N_2}{N_1} = \frac{G_1}{G_2}$$

EXAMPLE 3.10

A motor with a speed of 720 rev/min provides a torque of 50 Nm, and is fitted with a gearwheel having 30 teeth. This wheel is meshed with a second gearwheel, which has 12 teeth. Calculate the speed of the second gearwheel, and the torque it provides.

$$N_2 = \frac{N_1 G_1}{G_2} = 720 \times \frac{30}{12} \text{ revolutions per minute} = 1800 \text{ rev/min}$$

$$T_2 = \frac{T_1 G_2}{G_1} = 50 \times \frac{12}{30} \text{ newton metres} = 20 \text{ Nm}$$

Chains

A chain coupled to two or more rotating sprockets can be considered as a pulley-and-belt system, and follows the same rules. The chain is more expensive and is noisier than the belt, but cannot slip. It is used where a positive drive is needed.

Rotating shafts

As well as the drives already considered, the simple rotating shaft is a common mechanical power transmitter. Without belts, gears or chains, there is no change in speed or in torque with a shaft drive.

3.5 THE PARALLELOGRAM AND TRIANGLE OF FORCES

Scalars and vectors

Mechanical-engineering quantities can be classified as either scalars or vectors. Those which can be measured in terms of magnitude only are called **scalar** quantities. Examples of scalar quantities are length, mass and power.

Other quantities can only be completely described if reference is made to direction as well as to magnitude, and these are called **vector** quantities. For instance, the position of an aircraft after a given time can only be forecast accurately if not only its starting position and speed, but also its direction, is known.

52 Mechanics

A force is a vector quantity and is often represented by a line called a **vector**. The length of the vector, drawn to a suitable scale, represents the magnitude of the force, and the direction of the line is the same as that of the force it represents. An arrowhead is added to show the sense (push or pull) of the force, which is assumed to act at a point at one end of the vector.

Equilibrium

A body is said to be in a state of equilibrium when it is at rest or in a state of uniform motion. For instance, a lighting fitting suspended from a chain attached to a ceiling is in equilibrium because it does not move. The downward acting force due to the weight of the fitting is opposed by the equal upward supporting force of the chain. If the weight of the fitting is too great for its fixings to bear, the downward force exceeds the upward force, equilibrium is lost, and the fitting will fall.

Resultant of forces

It often happens that a body is acted on simultaneously by a number of forces. For simplicity, we can assume that these forces are removed and replaced by one force, called the **resultant**, which will have exactly the same effect on the body as the forces which it replaces. The resultant of two forces can be found by completing the **parallelogram of forces** as shown in Fig. 3.7. Two forces, P and Q, represented by the vectors OA and OB, have the resultant R, represented by the vector OC. If a resultant for more than two forces is required, a similar procedure is adopted, taking the resultant of two vectors at a time until the system is resolved into one vector only.

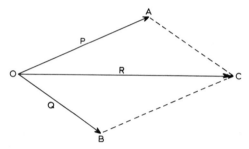

Fig. 3.7 Parallelogram of forces. R is resultant of forces P and Q

It is often necessary to find the force which will just balance all the other forces acting on a body. This is called the **equilibrant**, and is equal in magnitude and opposite in direction to the resultant of the forces.

EXAMPLE 3.11

A pole supporting overhead cables is situated so that the cables meet it at a right angle. The cables in one direction exert a pull of 4000 N on the pole, and those in the other direction a pull of 3000 N. Find the horizontal pull on a stay wire, and the direction in which it must be fixed so as to balance the horizontal forces exerted by the cables on the pole.

The arrangement of the system is shown in Fig. 3.8. O represents the pole, and vectors OA and OB represent the forces of 4000 N and 3000 N respectively. These vectors are drawn to scale and at right angles to each other. The parallelogram of forces is now

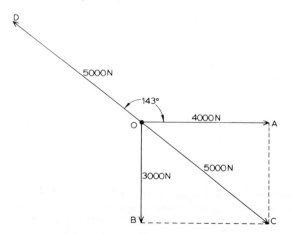

Fig. 3.8 Vector diagram for example 3.11

completed, the resultant vector OC representing a force of 5000 N in the direction shown.

The stay wire must be the equilibrant of the resultant OC, and is represented by the equal and opposite vector OD. Thus the horizontal pull on the stay wire will be 5000 N, and its direction, measured on the vector diagram with a protractor, will be approximately 143° from the 4000 N cables in an anticlockwise direction.

Triangle of forces

In Fig. 3.9a, the parallelogram of forces is used to find the resultant C and the equilibrant D of two forces A and B. Fig. 3.9b shows a part of the parallelogram used to find the resultant C, and illustrates a useful saving in space which could lead to a larger scale and improved accuracy. Fig. 3.9c is the same as Fig. 3.9b, but with the equilibrant D substituted for the resultant C. The three forces shown form the three sides of the triangle of forces.

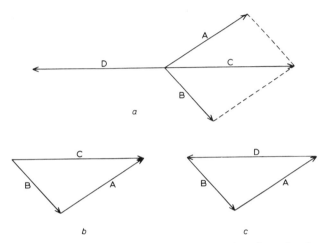

Fig. 3.9   a Parallelogram of forces used to find resultant C and equilibrant D of two forces A and B
         b Part of parallelogram of forces used to find resultant of same two forces
         c Triangle of forces used to find equilibrant of same two forces

EXAMPLE 3.12

A fluorescent fitting weighing 140 N is to be suspended from a single support chain connected to two sling chains taken to the fitting ends as shown in Fig. 3.10a. Find the tension in each sling chain, if each makes an angle of 30° with the fitting.

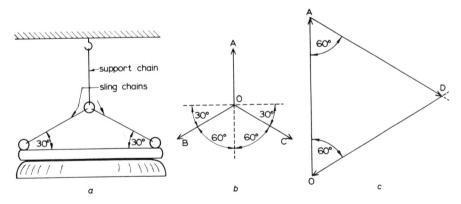

Fig. 3.10   a Fluorescent fitting supported from sling chains
          b Space diagram for arrangement of Fig. 3.10a
          c Triangle of forces for arrangement of Fig. 3.10a

A simple figure, often called a space diagram (Fig. 3.10b) will help us to understand the problem. Since the fitting is in a state of equilibrium, so must be the three forces in the space diagram, where OA is the tension in the support chain, and OB and OC are the tensions in the two sling chains. Clearly, the tension in the support chain must be equal to the weight of the fitting, and will be 140 N.

54   Mechanics

The triangle of forces is now drawn. First OA, to a scale length equivalent to 140 N, is drawn as shown in Fig. 3.10c. Next, lines are drawn from points O and A, each making an angle of 60° with OA and meeting at point D to complete the triangle. These lines represent the tensions in the sling chains, and, when measured, are found to have a length which is equivalent to 140 N.

Although the sling chains support the weight between them, each has the same tension as the support chain. Failure to appreciate the fact that the tension in the sling chains increases as the angle they make with the fitting decreases has been responsible for a number of accidents. Here the sling angles are equal; if they were unequal, the chain tensions would be different.

3.6 SUMMARY OF FORMULAS FOR CHAPTER 3

A mass of 1 kg experiences a force due to gravity of 9·81 N.

$$\text{Pressure, N/m}^2 = \frac{\text{force, N}}{\text{effective area, m}^2}$$

$$\text{Density, kg/m}^3 = \frac{\text{mass, kg}}{\text{volume, m}^3}$$

$$\text{Torque, Nm} = \text{turning force, N} \times \text{radius of action, m}$$

$$\text{Work of energy, J} = \text{force required, N} \times \text{distance moved, m}$$

$$\text{Efficiency} = \frac{\text{output}}{\text{input}} = \frac{\text{output}}{\text{output} + \text{losses}} = \frac{\text{input} - \text{losses}}{\text{input}}$$

$$\text{Power, W} = \frac{\text{work or energy, J}}{\text{time, s}}$$

$$\text{Mechanical advantage} = \frac{\text{load, N}}{\text{effort, N}}$$

$$\text{Velocity ratio} = \frac{\text{distance moved by effort, m}}{\text{distance moved by load, m}}$$

$$\text{Efficiency} = \frac{\text{mechanical advantage}}{\text{velocity ratio}}$$

$$\frac{N_1}{N_2} = \frac{D_2}{D_1} = \frac{T_2}{T_1}$$

where N_1 is the speed of the first pulley or gear, rev/min
N_2 is the speed of the second pulley or gear, rev/min
D_1 is the diameter of the first pulley or gear, m
D_2 is the diameter of the second pulley or gear, m
T_1 is the torque provided to, or by, the first pulley or gear, Nm
T_2 is the torque provided to, or by, the second pulley or gear, Nm

3.7 EXERCISES

1 A sling is marked as having a safe working load of 2000 kg. What weight will it support safely?

2 A machine requires a force of 500 N to lift it. What is its mass?

3 A masonry drill has an effective surface area of 50 mm² and, for best rate of penetration, must be operated at a pressure of 4 N/mm². What force should be applied to the drill?

4 An apprentice weighs 480 N. What pressure will be exerted on a plasterboard ceiling if he stands between the joists
   (a) on one foot which has an area of contact with the ceiling of 8000 mm²
   (b) on a rigid plank 300 mm wide and 2 m long?

5 A turning moment of 18 Nm is required to tighten a nut on a busbar clamp. What maximum force must be applied to a spanner of effective length 0·27 m if the nut is not to be overtightened?

6 An electrician exerts a turning force of 50 N on each of two handles of a set of stocks and dies. The effective length of each handle is 200 mm. What total turning moment is applied to the dies?

7   How much energy must be expended to raise a bundle of conduit weighing 600 N from a floor to a scaffolding 7 m above it? If the task takes two minutes to complete, what average power is used?

8   A motor has timbers bolted to it for protection, and requires a horizontal force of 450 N to move it over a level floor. How much work is done in moving it 50 m?

9   The petrol engine of a builder's hoist is to be replaced with an electric motor. What should be the rating of the motor if it must be capable of lifting 2400 N through 32 m in 24 seconds, the hoist gear being 80% efficient?

10 The pivot of a hydraulic pump is at one end of the handle, which is 1·2 m long, and is 0·2 m from the attachment to the pump rod. The pump piston has an effective surface area of 200 mm^2. If the handle of the pump is pushed down by a force of 80 N, calculate (*a*) the force on the pump rod and (*b*) the pressure on the pump piston.

11 What must be the minimum weight of a man who is able to lift a load of 5600 N using a crowbar 2 m long, with the fulcrum 0·2 m from the load end, and what is the mechanical advantage?

12 A block and tackle of eight pulleys is to be used to lift a cubicle switch panel weighing 2400 N. The lifting system is 80% efficient. Calculate the effort required.

13 A screw jack has a velocity ratio of 112 and an efficiency of 50%. What force must be applied to it just to lift a drum of cable weighing 8960 N?

14 A load pulley is 0·5 m in diameter, turns at 200 rev/min, and has a torque of 800 Nm. If the motor speed is 720 rev/min, calculate the size of the motor pulley and the torque it provides.

15 A load shaft is required to turn at 725 rev/min and to provide a torque of 1000 Nm. If the pulley on the 1450 rev/min motor is 250 mm in diameter, calculate the diameter of the load pulley and the torque provided by the motor.

16 A motor providing a torque of 100 Nm at 2900 rev/min is fitted with a gearwheel having 20 teeth. A drive is to be provided to a shaft having a gearwheel in mesh with that on the motor. If the shaft must provide a torque of 650 Nm, how many teeth has the driven gear, and what is its speed?

17 Two horizontal cables are attached to a pole. The first exerts a force of 2000 N, and the second a force of 3200 N making an angle of 120° with the first. Find the resultant pull on the pole, and its direction with respect to the first cable.

18 Two runs of cable exert forces of 6000 N and 8000 N, respectively, on an overhead line tower to which they are attached at right angles. What must be the direction of a stay, and what will be the horizontal force it supports, if there is to be no resultant horizontal force on the tower?

19 A set of pulley blocks, having four pulleys in the top block and three pulleys in the lower block, is fixed to a ceiling beam in a workshop. It is to be used of lift a motor weighing 2520 N from its bedplate. The efficiency of the tackle is 60%. Calculate the pull required on the free end of the rope to raise the motor. Make a diagrammatic sketch of the tackle in use.

20 A motor armature weighing 2000 N is freely suspended from a crane hook by means of a double sling with 1 m chains. The motor shaft is horizontal, and the slings are attached to the motor shaft 1 m apart.
    (*a*) Make a diagram showing the arrangement.
    (*b*) Determine the tension in the sling.
    (*c*) What would be the tension if the chains were 0·835 m instead of 1 m long?

Chapter 4

Heat

4.1 HEAT

In early childhood, we become familiar with the sensations of cold and warmth, and are able to distinguish between them. We learn to estimate the degree of hotness or coldness of a body, which is known as its **temperature.**

Heat is a form of energy. Heat added to a body makes it hotter, and heat taken away from a body makes it colder. It is possible, for instance, to increase the heat energy contained in a piece of metal, and hence to increase its temperature, by doing work such as cutting, bending or hammering it. Again, if work is done against friction, or if a fire is lit beneath the metal, it will become hotter as it absorbs part of the energy made available to it. Thus, when heat energy is produced, energy in some other form is expended.

Most of the losses of energy which occur in machines appear as heat, which is usually lost to the process concerned, although not destroyed.

4.2 TEMPERATURE

Temperature is a measure of the degree of hotness or coldness of a body. We are often concerned with the accurate measurement of temperature, for which an instrument called a **thermometer** is used. Many thermometers rely for their operation on the property of many materials to expand as their temperature increases (see Section 4.6). Glass-bulb thermometers consist of a mercury- or alcohol-filled bulb attached to the bottom of a fine glass tube. The liquid rises up the tube as it expands owing to increasing temperature, which can be read off from a scale beside the tube. Increasing use is being made of thermometers which consist of a long metal strip in the form of a coil. One end is securely fixed, and movement of the other end as the strip expands moves a pointer over a scale.

Increasing use is being made of electronic thermometers. These instruments have a thermistor (see section 2.6) enclosed in a probe, which is placed at the point whose temperature is required. The probe is connected to the thermometer, which measures the resistance of the thermistor and calculates its temperature. This is displayed, usually as a digital readout. As well as being more accurate and more easily read than the other types of thermometer, the electronic type has the advantage that the protected probe can often be put in positions not accessible to normal thermometers. For example, the probe may be distant from the display, it may be driven into frozen food, it may be lowered deep into liquid, and so on.

Temperature scales

Before any scale of measurement can be decided, two fixed points are necessary. For instance, in length measurement, the distance between two marks on a metal bar can indicate a standard length. For temperature scales, the upper fixed point is the temperature of steam from boiling pure water, and the lower fixed point is the temperature at which ice just melts, both measured at normal atmospheric pressure. Between these fixed points, the scale may be subdivided into any suitable number of parts. **The Celsius or centigrade scale** takes the lower fixed point as zero degree Celsius (written $0°C$) and the upper point as $100°C$.

The kelvin scale takes as its zero the lowest possible temperature, (called the absolute zero) and written as $0\,K$. A temperature change of one kelvin is the same as a change of one degree Celsius, so if we say that $0\,K = -273°C$, it follows that $0°C = 273\,K$. Thus, any temperature expressed in degrees Celsius can be given in kelvins by the addition of 273; or any temperature given in kelvins can be converted to degrees Celsius by the subtraction of 273.

EXAMPLE 4.1

(a) Convert to kelvins (i) $20°C$ (ii) $230°C$ (iii) $-40°C$.
(b) Convert to degrees Celsius (i) $265\,K$ (ii) $300\,K$ (iii) $1200\,K$.

a (i) $20°C = (20 + 273)\,\text{kelvins} = 293\,K$
 (ii) $230°C = (230 + 273)\,\text{kelvins} = 503\,K$
 (iii) $-40°C = (-40 + 273)\,\text{kelvins} = 233\,K$

b (i) $265\,K = (265 - 273)\,\text{degrees Celsius} = -8°C$
 (ii) $300\,K = (300 - 273)\,\text{degrees Celsius} = 27°C$
 (iii) $1200\,K = (1200 - 273)\,\text{degrees Celsius} = 927°C$

Strictly speaking, the kelvin scale is that used in SI units. However, the Celsius scale has more convenient numbers for everyday temperatures, and its widespread use is likely to continue.

4.3 HEAT UNITS

Heat is a form of energy, so the unit of heat is the same as that for energy, the joule. It can be shown experimentally, but not proved mathematically, that to heat 1 l of water (mass 1 kg) through 1°C, 4187 J of heat energy must be given to it.

This figure 4187 is called the **specific heat** of water. Different substances have different specific heats, but, in every case, the specific heat is the heat energy in joules needed to raise the temperature of one kilogram of the substance by one kelvin. The temperature of a body depends not only on the heat contained in the body, but also on its mass and its capacity to absorb heat. For instance, a small bowl of very hot water will be at a higher temperature than a bath full of cold water, but may contain less heat energy owing to its smaller mass. Again, 1 kg of brass will increase in temperature by 1 K for the addition of 376·8 J, whereas the same mass of water requires 4187 J for this temperature change. Table 6 shows the specific heats of a few common materials.

Table 6 Specific heats

Substance	Specific heat, J/kg K	Substance	Specific heat, J/kg K
Water	4187	Copper	397
Air	1010	Brass	376
Aluminium	915	Lead	129
Iron	497		

Thus we can see that the quantity of energy which must be added to a body to raise its temperature (or which must be removed from a body to lower its temperature) depends on its mass, its specific heat, and the temperature change involved.

Heat energy for temperature change, J = mass, kg × temperature change, K × specific heat, J/kg K

EXAMPLE 4.2

An immersion heater is required to raise the temperature of 50 l of water from 10°C to 85°C. If no heat is lost, find the energy required.

$$\text{Energy required} = \text{mass} \times \text{temperature change} \times \text{specific heat}$$

From Table 6, the specific heat of water is 4187 J/kg K. 1 l of water has a mass of 1 kg.

$$\text{Energy required} = 50 \times (85 - 10) \times 4187 \text{ J}$$
$$= 50 \times 75 \times 4187 \text{ J}$$
$$= 15\,700\,000 \text{ J or } 15 \cdot 7 \text{ MJ}$$

In practice, heat will be lost through the container during heating, so the efficiency of the system must be taken into account.

EXAMPLE 4.3

A tank containing 150 l of water is to be heated from 15°C to 75°C. If 25% of the heat provided is lost through the tank, how much energy must be supplied?

From Table 6, the specific heat of water is 4187 J/kg K. If 25% of the heat is lost, 75% is retained and the system is 75% efficient.

Energy required at 100% efficiency $= 150 \times (75 - 15) \times 4187$ joules

The energy required at 75% efficiency will be greater, and will be

$$150 \times (75 - 15) \times 4187 \times \frac{100}{75} \text{ joules}$$
$$= 150 \times 60 \times 4187 \times \frac{100}{75} \text{ joules}$$
$$= 50\,240\,000 \text{ J or } 50 \cdot 24 \text{ MJ}$$

60   Heat

EXAMPLE 4.4

2 kg of iron are placed in a furnace at an initial temperature of 15°C, and heat energy of 3·976 MJ is provided to the furnace. If one quarter of the heat provided is received by the iron, what will be its final temperature?

From Table 6, the specific heat of iron is 497 J/kg K.

If the total heat provided is 3·976 MJ, and one-quarter of this reaches the iron, the heat absorbed by the iron is $\frac{3 \cdot 976}{4}$ megajoules, or 0·994 MJ.

$$\text{Heat provided} = \text{mass} \times \text{temperature change} \times \text{specific heat}$$

$$\text{temperature change} = \frac{\text{heat provided}}{\text{mass} \times \text{specific heat}}$$

$$= \frac{0 \cdot 994 \times 10^6}{2 \times 497} \text{ K}$$

$$= \frac{0 \cdot 994 \times 10^6}{994} \text{ K}$$

$$= 1000 \text{ K}$$

If the initial temperature is 15°C and increases by 1000 K, final temperature will be 1015°C.

4.4 HEATING TIME AND POWER

We have already seen that a rate of doing work of one joule per second is a power of one watt, or

$$\text{watts} \times \text{seconds} = \text{joules}$$

It is common for the kilowatt hour (kWh) to be used as a larger and more convenient unit than the joule. A kilowatt hour is the energy used when a power of 1000 watts applies for one hour.

$$1 \text{ kWh} = 1000 \text{ W} \times 60 \text{ min} \times 60 \text{ s}$$

$$= 3\,600\,000 \text{ Ws or J},$$

$$\text{or } 1 \text{ kWh} = 3 \cdot 6 \text{ MJ}$$

The rating of electric heaters is usually in watts or in kilowatts, and the time relationship of this power with heating energy enables us to calculate the time taken for a heater of a given rating to provide a given quantity of heat. Alternatively, the rating necessary to provide a given quantity of heat in a given time can be found. These calculations are illustrated in the following examples.

Although electric heating is probably the most efficient of the heating methods, some heat may be lost, and this must be taken into account.

EXAMPLE 4.5

How long will a 3 kW immersion heater take to raise the temperature of 30 l of water from 10°C to 85°C? Assume that the process is 90% efficient.

$$\text{Heat required} = 30 \times (85 - 10) \times 4187 \times \frac{100}{90} \text{ joules}$$

$$= 30 \times 75 \times 4187 \times \frac{100}{90} \text{ joules}$$

$$= 10\,470\,000 \text{ J} \quad \text{or} \quad 10 \cdot 47 \text{ MJ}$$

A 3 kW heater gives 3000 W, or 3000 J/s.

$$\text{Therefore time required} = \frac{10\,470\,000}{3000} \text{ seconds}$$

$$= 3490 \text{ s}$$

$$= \frac{3490}{60} \text{ minutes}$$

$$= 58 \text{ min } 10 \text{ s}$$

EXAMPLE 4.6

A jointer's pot contains 25 kg of lead, and is heated by a gas flame providing a power of 2 kW to the pot. If the initial temperature of the lead is 15°C, how hot will it be after 5 min?

From Table 6, the specific heat of lead is 129 J/kg K. We can assume 100% efficiency, since 2 kW is provided to the pot. The heat lost will be several times this figure for a heater of this type.

$$\text{Heat provided} = 2 \text{ kW for 5 min}$$
$$= 2000 \times 5 \times 60 \text{ joules}$$
$$= 600\,000 \text{ J}$$
$$\text{heat used} = \text{mass} \times \text{temperature change} \times \text{specific heat}$$

Since heat provided = heat used,

$$600\,000 = 25 \times \text{temperature change} \times 129$$
$$\text{temperature change} = \frac{600\,000}{25 \times 129}$$
$$= 186 \text{ K}$$

Since the initial temperature is 15°C and the increase is 186 K, final temperature is 201°C.

EXAMPLE 4.7

A 3 kW water heater is of the instantaneous type, which heats water as it flows over its element. Such a heater can be assumed to be 100% efficient. What will be the increase in water temperature from inlet to outlet if the rate of flow is 1 l/min?

$$\text{Since 3 kW} = 3000 \text{ J/s},$$
$$\text{heat provided in 1 min} = 3000 \times 60 \text{ J}$$
$$= 180\,000 \text{ J}$$
$$\text{Heat provided} = \text{mass} \times \text{temperature change} \times \text{specific heat}.$$
$$180\,000 = 1 \times \text{temperature change} \times 4187 \text{ J}$$
$$\text{Temperature change} = \frac{180\,000}{1 \times 4187} \text{ kelvins}$$
$$= 43 \text{ K}$$

4.5 HEAT TRANSMISSION

In Chapter 3, we said that some of the heat produced by a machine is 'lost' to the atmosphere. This indicates clearly that heat must be able to move from the point at which it is generated. Transmission of heat is an essential part of many engineering processes. For instance, heat must be transferred from the bit of a soldering iron to the solder and to the surfaces to be joined. Again, heat is given out from the element of an electric fire; if it were not, the continuous input of energy by means of the electric current would increase the temperature of the element until it melted.

The amount of heat transmitted depends on the difference in temperature between the heat source and the places to which the heat moves. When an electric fire is first switched on, very little heat is transmitted, because the element temperature is low. As time goes by, electrical energy is still fed into the element and appears as heat, raising the temperature and increasing the heat transmitted. In due course, the element becomes hot enough to transmit the same amount of heat energy as it receives. In this condition of **heat balance**, the temperature of the element remains constant. This heat balance applies to all devices in which heat is generated, the final steady temperature depending on the energy input and the ease with which heat can be transmitted.

In practice, heat is transmitted by three separate processes which can occur individually or in combination. These processes are called conduction, convection and radiation.

Conduction

If one end of a metal bar is heated, the other end also becomes hot. Heat has been **conducted** along the length of the bar from the high-temperature end to that at the lower temperature, the heat energy trying to distribute itself evenly throughout the bar. If we attempt a similar experiment with wood, the heated end can burn without the colder end increasing in temperature to any extent. Thus metal is said to be a good conductor of heat, whereas wood is a very poor conductor.

Metals are usually good conductors of heat, and wood, plastic and similar materials are not. For instance, a soldering iron will have a copper bit to allow maximum heat conduction to the work, but will have a wood or plastic handle to prevent it becoming too hot to hold.

Most liquids and gases are poor conductors of heat. Because of this, materials which trap air in pockets, such as felt, glass fibre and cork, are used to prevent the conduction of heat, and are called **heat insulators**. An application of such materials, often called **lagging**, is used to prevent heat escape from ovens, hot-water tanks and the like. The same sort of lagging may be used to prevent heat entering a refrigerator.

Convection

Transmission of heat by convection takes place in fluids, i.e. in liquids and gases. If a given volume of a liquid or a gas is heated, it expands if free to do so, and so the same volume weighs less than that of the unheated fluid. We can say that the density has become lower. It will tend to rise, its place being taken by cool fluid, which will also rise when heated, so that a steady upward current of warm fluid results.

This principle is widely used in some forms of air heaters, called convectors, which draw in cool air at the bottom and expel hot air from the top.

The same sort of process is used in some types of central-heating system, where water circulates through the radiators solely as a result of convection currents. (In large installations, or those using small-bore piping, the natural circulation is assisted by a pump.)

The contents of a hot-water tank also circulate owing to convection currents. Water is, however, a very bad conductor of heat, and the water below the level of an immersion heater remains cold and is not displaced by convection currents. For this reason, two heaters are often fitted to hot water cylinders. The upper heater maintains about one third of the water hot for normal uses, the lower heater being switched on when large amounts of hot water are required.

Radiation

Most of the energy reaching the Earth does so in the form of heat radiated from the Sun. Unlike conduction and convection, which can occur only in material substances, radiation of heat from one body to another does not require any connecting medium between the bodies. Heat is radiated freely through space in much the same way as light. All bodies radiate energy all the time, the rate of radiation depending mainly on the temperature of the body and its surroundings. Thus a cold body surrounded by warm objects will radiate less heat than it receives, and will tend to become warmer.

Radiant heat behaves in a very similar manner to light, and is reflected from bright surfaces. A highly polished reflector is fitted to a fire with a rod-type element, and this reflector beams the heat emitted by the element in a particular direction. Fires of this sort are often called radiators, although much of their energy is given off in the form of convection, and a little by conduction.

A body with a dark, matt surface tends to absorb heat instead of reflecting it. If the reflector of a radiator were to be painted black, it would absorb heat quickly and become excessively hot. Bright surfaces do not emit (as opposed to reflect) radiation so readily as dull ones. For this reason, appliances such as electric kettles are often chromium plated to reduce heat losses and thus improve efficiency.

4.6 CHANGE OF DIMENSIONS WITH TEMPERATURE

Most materials increase slightly in dimensions when their temperature increases. For instance, overhead lines sag more in the summer than in the winter, and a long straight conduit run may buckle in very hot weather if expansion bends are not fitted. Similarly, dimensions often decrease when temperature decreases. An overhead cable that is erected in hot weather and strained tightly will contract in cold weather; the extra stress may then result in stretching or even breaking.

Although expansion with increasing temperature is often a nuisance, it can be applied with advantage when used to control temperature. Some metals expand more than others when heated through the same temperature range. If two strips, one of each of two metals with different rates of expansion, are riveted together, the **bimetal strip** so formed will bend when heated. If a set of contacts are operated by the strip as it bends, the device can be made to control temperature, and is called a **thermostat**. Another type of thermostat based on the same principle is called a rod-type thermostat. A rod of material, selected for its small increase in length when heated, is mounted within a tube of brass, the rod and the tube being welded together at one end. Changes in the temperature of the device result in differing changes in length of the rod and the tube, thus operating a switch. The slow break resulting from the slow rate of differential expansion will give rise to arcing at the contacts of a directly operated switch. Permanent-magnet systems and flexed springs are often used to give a quick make-and-break action to the switch (Figs. 4.1 and 4.2).

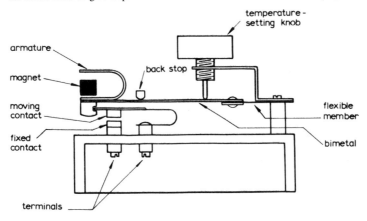

Fig. 4.1   Air thermostat

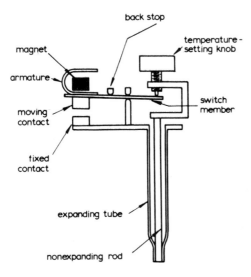

Fig. 4.2   Rod-type immersion thermostat

4.7   SUMMARY OF FORMULAS FOR CHAPTER 4

$$\text{Temperature in kelvins} = \text{temperature in degrees Celsius} + 273$$

$$\text{Temperature in degrees Celsius} = \text{temperature in kelvins} - 273$$

$$\text{Heat energy for temperature change, J} = \text{mass, kg} \times \text{temperature change, K} \times \text{specific heat, J/kg K}$$

$$\text{Energy, J} = \text{power, W} \times \text{time, s}$$

4.8 EXERCISES

1. Express the following temperatures in kelvins:
   (a) 60°C,
   (b) −75°C,
   (c) 1000°C

2. Express the following temperatures in degrees Celsius:
   (a) 320 K
   (b) 1500 K
   (c) 240 K

3. A small storage heater contains 8 l of water at a temperature of 10°C. How much heat energy must be provided to raise the water temperature to 90°C? The specific heat of water is 4187 J/kg K.

4. How much heat energy must be supplied to 20 kg of brass to increase its temperature by 500 K? Give your answer in joules and in kilowatt hours. The specific heat of brass is 376 J/kg K.

5. A hall measures 10 m by 30 m, by 5 m high. How much heat energy will be required to raise the temperature of the air it contains from 5°C to 22°C? 1 m^3 of air has a mass of 1·26 kg. The specific heat of air is 1010 J/kg K. Express your answer in kilowatt hours.

6. Calculate the amount of heat in joules required to raise the temperature of 100 kg of aluminium from 10°C to 710°C. The specific heat of aluminium is 915 J/kg K.

7. 40 l of water is heated from 8°C to 78°C, the efficiency of the operation being 70%. How much heat is required? The specific heat of water is 4187 J/kg K.

8. The rate of flow of water through a water-cooled motor is 0·2 l/s, and inlet and outlet temperatures are 10°C and 20°C, respectively. At what rate is heat being removed from the motor? The specific heat of water is 4187 J/kg K.

9. An electric-arc furnace is used to raise the temperature of 4000 kg of iron from 12°C to 812°C, the overall efficiency of the furnace being 40%. What energy input in kilowatt hours is required? The specific heat of iron is 497 J/kg K.

10. An electric water heater is 80% efficient and consumes energy at the rate of 2000 J/s. If the heater initially contains 10 l of water at 12°C, what will be the water temperature after 10 min of heating? The specific heat of water is 4187 J/kg K.

11. The input power to a furnace for heating copper is 10 kW, and the furnace is 39·7% efficient. By how much will the temperature of 150 kg of copper have increased after 30 min? The specific heat of copper is 397 J/kg K.

12. A 2 kW heater is switched on in a room 5 m square and 3 m high, when the air temperature is 70°C. If 70% of the heat provided is lost, what will be the air temperature after 20 min? 1 m^3 of air has a mass of 1·26 kg, and the specific heat of air is 1010 J/kg °C.

13. How long will it take a 1·5 kW heater to raise the temperature of 10 l of water by 60°C if the heater is 100% efficient? The specific heat of water is 4187 J/kg K.

14. An instantaneous-type water heater is required to provide a continuous flow of water of 2 l/min at a temperature of 85°C with an inlet water temperature of 10°C. What must the the electrical loading in kilowatts if the heater is assumed to be 100% efficient? The specific heat of water is 4187 J/kg K.

15. A furnace for lead casting contains 120 kg of lead, and is heated by a 6 kW element. The initial temperature of the lead is 20°C. What temperature will the lead reach after 1 h if the furnace is 25% efficient? The specific heat of lead is 129 J/kg K.

Chapter 5
Electrical power and energy

5.1 UNITS OF ELECTRICAL POWER AND ENERGY

We found in Chapter 1 that a p.d. of one volt exists between two points if one joule of energy is expended in moving a coulomb of electricity between them. Thus

$$\text{joules} = \text{coulombs} \times \text{volts}$$

Since the coulomb is current × time

$$\text{joules} = \text{amperes} \times \text{volts} \times \text{seconds, or } W = IVt$$

Power, as shown in Chapter 3, is the rate of doing work. The mechanical unit of power is the watt (W), and this is also the unit for electrical power. There is no reason for different units of power to exist, because the work done at a given rate can be the same whether provided by an electric motor, or a mechanical method such as a diesel engine.

The watt is defined as a rate of doing work of one joule per second.

Therefore $$P = \frac{\text{joules}}{\text{seconds}} = \frac{IVt}{t} = IV$$

In other words, the power expended in watts at any instant in a given circuit is given by the instantaneous circuit current in amperes times the instantaneous circuit p.d. in volts, or

$$P = IV \text{ for d.c. circuits}$$

It is often useful to be able to express the power P expended in a circuit in terms of its resistance, together with either current or p.d.

$$P = IV, \text{ but } V = IR, \text{ so } P = I(IR) = I^2R$$

Also $$P = IV, \text{ but } I = \frac{V}{R}, \text{ so } P = \frac{VV}{R} = \frac{V^2}{R}$$

It should be noted that all of these expressions are not always true for a.c. circuits.

EXAMPLE 5.1

A 100 Ω resistor is connected to a 10 V d.c. supply. What power is dissipated in it?

$$P = \frac{V^2}{R}$$
$$= \frac{10 \times 10}{100} \text{ watt}$$
$$= 1 \text{ W}$$

or
$$I = \frac{V}{R}$$
$$= \frac{10}{100} \text{ ampere}$$
$$= 0.1 \text{ A}$$
$$P = IV$$
$$= 0.1 \times 10 \text{ watt}$$
$$= 1 \text{ W}$$

or
$$P = I^2R$$
$$= 0.1 \times 0.1 \times 100 \text{ watt}$$
$$= 1 \text{ W}$$

EXAMPLE 5.2

What is the hot resistance of a 240 V, 100 W lamp?

$$P = \frac{V^2}{R}, \text{ so } R = \frac{V^2}{P}$$

$$= \frac{240 \times 240}{100} \text{ ohms}$$

$$= 576 \, \Omega$$

Since watts = joules per second, it follows that the joule, the unit of energy, is the watt second.

EXAMPLE 5.3

How much energy is supplied to a 100 Ω resistor which is connected to a 150 V supply for 1 h?

$$P = \frac{V^2}{R}$$

$$= \frac{150 \times 150}{100} \text{ watts}$$

$$= 225 \, \text{W}$$

$$W = Pt$$

$$= 225 \times 1 \times 60 \times 60 \text{ joules}$$

$$= 810\,000 \, \text{J}$$

Note: Time is expressed in seconds.

The joule is too small for the measurement of the amounts of energy in common use, so the **kilowatt hour** (kWh) is the unit for many practical and commercial purposes. The kilowatt hour is the energy used when a power of one kilowatt (1000 W) is used for 1 h (3600 s) and is often referred to as the 'unit' of energy.

Since joules = watts × seconds,

$$1 \, \text{kWh} = 1000 \times 3600 \text{ joules} = 3\,600\,000 \, \text{J}$$

For example 5.3, the energy is therefore 0·225 kWh.

EXAMPLE 5.4

A d.c. motor takes 15 A from a 200 V supply, and is used for 40 min. What will this cost, if the tariff is 6 p/kWh?

$$\text{Energy} = \frac{V \times I}{1000} \times \frac{\text{minutes}}{60} \text{ kilowatt hours}$$

$$= \frac{15 \times 200}{1000} \times \frac{40}{60} \text{ kilowatt hours}$$

$$= 2 \, \text{kWh}$$

$$\text{Energy charge} = 2 \times 6 \text{ pence} = 12 \text{p}$$

Another multiple of the joule, used because the joule is small, is the megajoule (MJ). One megajoule is equal to one million joules.

$$1 \, \text{MJ} = 1\,000\,000 \, \text{J, or } 10^6 \, \text{J}$$

Since 1 kWh = 3·6 × 10⁶ joules, it follows that

$$1 \, \text{kWh} = 3 \cdot 6 \, \text{MJ}$$

EXAMPLE 5.5

An electric heater consumes 2·7 MJ when connected to its 240 V supply for 30 min. Find the power rating of the heater, and the current taken from the supply.

68   Electrical power and energy

$$\text{Power} = \frac{\text{energy}}{\text{time}}$$

Therefore
$$P = \frac{2 \cdot 7 \times 10^6}{30 \times 60} \text{ watts}$$

$$= 1500 \text{ W or } 1 \cdot 5 \text{ kW}$$

$$P = VI, \text{ so } I = \frac{P}{V}$$

$$I = \frac{1500}{240} \text{ amperes}$$

$$= 6 \cdot 25 \text{ A}$$

EXAMPLE 5.6

An electric fire has two elements, each of resistance 18 Ω, and is connected to the supply through a cable of resistance 2 Ω. The supply is of 200 V. Calculate the total power taken from the supply, and the total current.

(a) if one element is switched on
(b) if both elements are switched on.

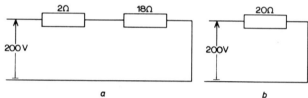

Fig. 5.1 Diagrams for example 5.6a

(a) The element resistance of 18 Ω is in series with the supply-cable resistance of 2 Ω, and can be represented as shown in Fig. 5.1a. Since these resistors are in series, they can be represented by one 20 Ω resistor, as shown in Fig. 5.1b.

Applying Ohm's law,
$$I = \frac{V}{R}$$

$$= \frac{200}{20} \text{ amperes}$$

$$= 10 \text{ A}$$

The power taken can be found by any of the three methods shown in example 5.1.

$$P = VI = 200 \times 10 \text{ watts} = 2000 \text{ W}$$

or
$$P = I^2 R = 10^2 \times 20 \text{ watts} = 2000 \text{ W}$$

or
$$P = \frac{V^2}{R} = \frac{200^2}{20} \text{ watts} = 2000 \text{ W}$$

(b) Since the two elements in parallel are connected in series to the supply cable, the circuit is effectively that shown in Fig. 5.2a. First, find the equivalent resistance of the two 18 Ω resistors in parallel.

$$\frac{1}{R} = \frac{1}{18} + \frac{1}{18} = \frac{2}{18}$$

Therefore
$$R = \frac{18}{2} \text{ ohms} = 9 \text{ Ω}$$

The circuit can then be redrawn as in Fig. 5.2b, and further simplified, since the 2 Ω and 9 Ω resistors are in series, to one 11 Ω resistor connected to the 200 V supply.

From Ohm's law,

$$I = \frac{V}{R}$$

$$= \frac{200}{11} \text{ amperes}$$

$$= 18 \cdot 2 \text{ A approximately.}$$

$$P = VI = 200 \times 18 \cdot 2 \text{ watts} = 3640 \text{ W}$$

Either of the methods $P = I^2 R$ or $P = V^2/R$ could be used as an alternative.

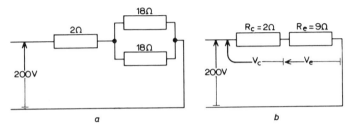

Fig. 5.2 Diagrams for examples 5.6b and 5.7

EXAMPLE 5.7

How much of the total power taken from the supply by the fire of example 5.7 is dissipated in the elements, and how much in the supply cable, if both elements are switched on?

If we use $P = V \times I$, the voltage used will have to be the voltage drop across the part of the circuit concerned.

Voltage drop in cable,

$$V_c = IR$$

$$= 18 \cdot 2 \times 2 \text{ volts}$$

$$= 36 \cdot 4 \text{ V}$$

Power dissipated in the cable will then be

$$P_c = V_c I = 36 \cdot 4 \times 18 \cdot 2 \text{ watts} = 662 \text{ W}$$

The power dissipated in the elements must then be the balance of the total power, so that

$$P_e = 3640 - 662 \text{ watts} = 2978 \text{ W}$$

Either of these answers could conveniently have been found by using the $P = I^2 R$ method.

$$P_c = I^2 R_c = 18 \cdot 2^2 \times 2 \text{ watts} = 662 \text{ W}$$

$$P_e = I^2 R_e = 18 \cdot 2^2 \times 9 \text{ watts} = 2978 \text{ W}$$

EXAMPLE 5.8

A house installation has the following circuit loads:

lighting: two 150 W lamps, six 100 W lamps and ten 60 W lamps
ring circuit: one 3 kW and one 2 kW heaters
water heating: one 3 kW immersion heater.

What is the current in each circuit if the supply is at 240 V? What size fuse should be used for each circuit?

Lighting:

Total load comprises 2×150 watts = 300 W
6×100 watts = 600 W
10×60 watts = 600 W
Total 1500 W

$$I = \frac{P}{V} = \frac{1500}{240} \text{amperes} = 6.25 \text{A}$$

A suitable fuse rating would be 10 A.

Ring circuit   Load comprises 2 kW + 3 kW = 5 kW

Therefore
$$I = \frac{P}{V} = \frac{5000}{240} \text{amperes} = 20.8 \text{A approximately.}$$

A suitable fuse rating would be 30 A.

Water heating: Load comprises 3 kW

$$I = \frac{P}{V} = \frac{3000}{240} \text{amperes} = 12.5 \text{A}$$

A suitable fuse rating would be 15 A.

EXAMPLE 5.9

6 Ω and 3 Ω resistors are connected in parallel, this group being connected in series with a 2 Ω resistor across a d.c. supply. If the power dissipated in the 6 Ω resistor is 24 W, what is the supply voltage?

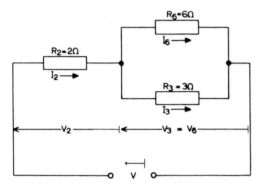

Fig. 5.3 Diagram for example 5.9

The circuit layout is shown in Fig. 5.3, which also gives a simple method of identifying the currents and voltages.

For the 6 Ω resistor, $P_6 = 24$ W and $R_6 = 6$ Ω

$$P_6 = I_6^2 R_6, \text{ so } I_6^2 = \frac{P_6}{R_6} = \frac{24}{6} \text{A}^2 = 4 \text{A}^2$$

Therefore $I_6 = \sqrt{I_6^2} = \sqrt{4} \text{A} = 2 \text{A}$

Thus $V_6 = I_6 R_6 = 2 \times 6 \text{volts} = 12 \text{V}$

Since the 6 Ω and 3 Ω resistors are in parallel,

$$V_6 = V_3 = 12 \text{V}$$

Therefore
$$I_3 = \frac{V_3}{R_3} = \frac{12}{3} \text{amperes} = 4 \text{A}.$$

But I_6 and I_3 are made up from current I_2, which splits after passing through R_2.

Thus $I_2 = I_6 + I_3 = 2 + 4 \text{amperes} = 6 \text{A}$

Therefore $V_2 = I_2 R_2 = 6 \times 2 \text{volts} = 12 \text{V}$

Supply voltage V is made up of V_2 in series with the p.d. across the parallel bank, V_6 or V_3.

Therefore $V = V_2 + V_6 = 12 + 12 \text{volts} = 24 \text{V}$

5.2 ELECTROMECHANICAL CONVERSIONS

In SI units, there is no difference between mechanical and electrical quantities. Thus the watt is the unit of power in all systems, and is always defined as a rate of doing work of one joule per second. Similarly, the joule is the unit of energy in both mechanical and electrical systems. In mechanical applications, the joule is defined as the work that is done when a force of one newton is moved through one metre. For electrical applications, the joule can be defined as the work that is done when one coulomb of electricity is moved through a potential difference of one volt.

Since the same units are used for both systems, conversions from mechanical to electrical forms, and vice versa, are simple, relying on the following relationships:

$$\text{joules} = \text{coulombs} \times \text{volts} = \text{metres} \times \text{newtons}$$

Since coulombs = amperes × seconds,

$$\text{watts} = \frac{\text{joules}}{\text{seconds}} = \text{amperes} \times \text{volts} = \frac{\text{metres} \times \text{newtons}}{\text{seconds}}$$

EXAMPLE 5.10

Calculate the power rating of a lift which can raise a mass of 800 kg through a height of 50 m in 98·1 s.

$$\text{Work done} = \text{force (newtons)} \times \text{distance (metres)}$$

1 kg requires a force of 9·81 N to lift it against gravity, so 800 kg require 800 × 9·81 N.

$$\text{Work done} = 800 \times 9 \cdot 81 \times 50 \, \text{J}$$

$$\text{Power (W)} = \frac{\text{work done, J}}{\text{time, s}}$$

$$= \frac{800 \times 9 \cdot 81 \times 50}{98 \cdot 1} \, \text{watts}$$

$$= 4000 \, \text{W or } 4 \, \text{kW}$$

EXAMPLE 5.11

Calculate the output and input powers of a motor driving the lift of example 5.7 if the lift gear is 80% efficient and the motor is 90% efficient.

$$\text{Efficiency} = \frac{\text{output}}{\text{input}}$$

Input to the lift (which is also the output of the motor) $= \dfrac{\text{output}}{\text{efficiency}}$

$$\text{Output of motor} = 4 \times \frac{100}{80} \, \text{kilowatts} = 5 \, \text{kW}$$

$$\text{Motor input} = \frac{\text{output}}{\text{efficiency}}$$

$$= 5 \times \frac{100}{90} \, \text{kilowatts}$$

$$= 5 \cdot 56 \, \text{kW}$$

EXAMPLE 5.12

If the motor of examples 5.10 and 5.11 is fed from a 220 V d.c. supply, what current will it take during the lift?

$$P = VI, \text{ so } I = \frac{P}{V}$$

$$I = \frac{5560}{220} \, \text{amperes} = 25 \cdot 2 \, \text{A}$$

EXAMPLE 5.13

A d.c. generator is to provide a maximum current of 500 A at 220 V, and is 0·88 efficient. What must be the kilowatt rating of the diesel engine driving it?

$$\text{Output} = 500 \times 220 \, \text{W}$$

$$\text{Input} = \frac{\text{output}}{\text{efficiency}}$$

$$= \frac{500 \times 220}{0 \cdot 88} \, \text{watts}$$

$$= \frac{500 \times 220}{0 \cdot 88 \times 1000} \, \text{kilowatts}$$

$$= 125 \, \text{kW}$$

5.3 ELECTRIC HEATING

We have already seen that, when an electric current passes through a resistor, power is produced. This power appears as heat in the resistor.

The production of heat due to dissipation of electrical energy is widespread, and examples are numerous. A few of the most common are listed below:

Filament lamps

A very thin tungsten wire is formed into a small coil and supported within a glass bulb. The passage of an electric current through this filament causes it to reach a temperature of 2500°C or more, so that it glows brightly. At these temperatures, the oxygen in the atmosphere would combine with the filament to cause failure, so all the air is removed from the glass bulb and replaced by a gas such as nitrogen, which does not react with the hot tungsten. The construction of a filament lamp is shown in Fig. 5.4.

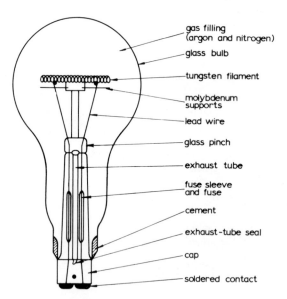

Fig. 5.4 Tungsten-filament lamp

Air heaters

The purpose of these devices is to warm the air in buildings, and thus make living and working more comfortable in cold weather. As shown in Chapter 4, the heat is transferred to the air from the resistive element by radiation and convection.

The radiant heater has an element that operates at a temperature high enough for it to glow red. Heat is radiated, and is often directed to the required area by the use of a polished reflector.

A convector heater usually has a low-temperature 'black heat' element. Air in contact with the element is warmed and becomes less dense, so that it rises and is replaced by colder air, which is warmed in turn. A continuous output of warm air results.

A forced convection, or fan-type, heater is shown in Fig. 5.5.

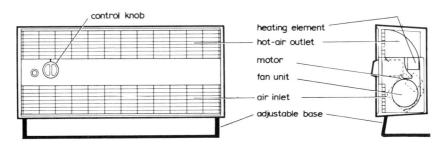

Fig. 5.5 Fan-type heater

Owing to the reduced cost of electricity taken only at times of low demand, storage heaters are widely used. The radiator type is heated by elements in fireclay blocks. The heat from these blocks is conducted through the lagging provided, which slows down the rate at which heat is released. The fan type storage heater has thicker lagging, so that little heat is lost through it. Heat is provided by passing air over the heated blocks by means of a fan. Such a heater is shown in Fig. 5.6. The solid floor of a building may also be used to store heat, heating cables being buried in it.

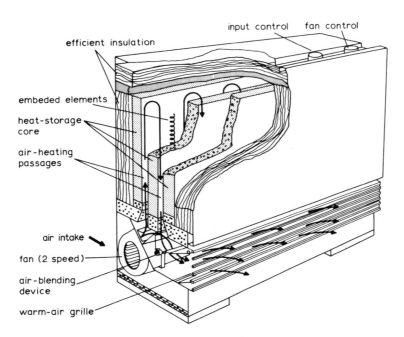

Fig. 5.6 Cut-away view of fan-type storage heater

Water heaters
An electric element is enclosed in a copper or nickel tube, and electrically insulated from it with a mineral insulation of magnesium oxide. The resulting heater is immersed in the water to be heated. Heat is transferred to the water by conduction, after which the water circulates naturally by a process of convection. There are many types of electric heater, one of the most common being shown in Fig. 5.7.

Other heating applications
There are very many more applications of electric heating in industry, in commerce and at home. For example, electric welding is widely used for fabrication, and electric furnaces are used to produce steel and other alloys.

In the home, the electric cooker, kettle, iron and firelighter are electric heaters that are often taken for granted.

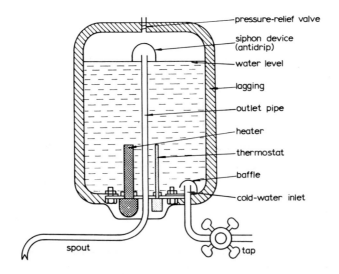

Fig. 5.7 Nonpressure or free-outlet-type water heater

5.4 SUMMARY OF FORMULAS FOR CHAPTER 5

$$P = \frac{W}{t} = IV = I^2R = \frac{V^2}{R}$$

where
- P = power dissipated, W
- W = energy expended, J
- t = time taken, s
- I = circuit current, A
- R = circuit resistance, Ω
- V = circuit potential difference, V

Note: $P = IV$ and $P = V^2/R$ are not always true for a.c. circuits.

$$1 \text{ kWh} = 3\,600\,000 \text{ J} = 3.6 \text{ MJ}$$

5.5 EXERCISES

1. An electric fire takes a current of 6·25 A from a 240 V supply. What power does it dissipate?

2. How much power will be dissipated in a 3 Ω resistor when it is connected across a 12 V battery?

3. A coil of resistance 5 Ω carries a current of 10 A. What power is dissipated in the coil?

4. A heating element takes a current of 5 A at 240 V. The power taken is . . . . (ULCI)

5. When a 5 kW heater is connected to a 250 V supply, it takes a current of . . . . (ULCI)

6. A 240 V electric kettle has a heating element of resistance 30 Ω, and a 200 V kettle has an element of resistance 20 Ω. Show, by calculation, which kettle has the greater rating. (ULCI)

7. A hall is lit by 20 lamps, each of which takes 0·5 A when connected to a 240 V supply. Calculate the total power supply to the lamps. (NCTEC)

8. An electric lamp is marked 150 W, 250 V. When it is connected to a 250 V supply, the current taken is . . . . (ULCI)

9. What is the resistance of an electric-fire element rated at 1 kW when carrying a current of 5 A?

10. When dissipating a power of 1 W, the p.d. across a carbon radio resistor is 100 V. What is the resistance?

11. What current will a 750 W fire element take from a 240 V supply?

12. How much current will a 10 Ω resistor be carrying if the power lost in the resistor is 1 kW?

13 A 3 kW heating element has a resistance of 30 Ω. For what supply voltage was the heater designed?

14 A 10 kΩ resistor in an electronic circuit has 0·5 W of power lost in it. What current is it carrying, and what is the p.d. across it?

15 How much energy will be consumed by the resistor of exercise 14 in 10 min?

16 An electric water heater must provide 4 MJ of energy to heat its contents in 50 min. What is the rating of the heating element?

17 An underfloor heating system has a loading of 20 kW. How much energy will it consume in 10 h?

18 A 10 Ω resistor carries a current of 5 A. How much energy will it dissipate as heat each minute?

19 A 2 Ω resistor liberates as heat 1440 J of energy in 20 s. What is the p.d. across the resistor?

20 A motor with an output of 8 kW and an efficiency of 80% runs for 8 h. Calculate the cost if electricity costs 6p per unit.

21 A generator in a motor car charges the battery at 20 A at 16 V. The generator efficiency is 43%. What power is absorbed in driving it?

22 The rating of the hotplate of an electric cooker is unknown. At 6 p.m., the meter reads 14 567 units. The cooker hotplate is at once switched on, together with a 1 kW fire and 500 W of lighting, and left on until 10 p.m., when the meter reads 14 581 units. What is the rating of the hotplate?

23 It costs 40p to run a motor for 2 h, the tariff being 6·67p per unit. If the motor is 90% efficient, what is its output?

24 An electric water heater provides 48 MJ in 40 min. What is the rating of the element, and how much current does it take from a 240 V supply?

25 A workshop supplied at 250 V has two motor drives at 1 kW each, a fan which takes 1 A, two 100 W lamps, a 2 kW radiator and a battery charger which takes 50 W. Calculate the total load in kilowatts and the weekly cost at 6p per unit if the load is on for 40 h a week.

26 A resistor of unknown value is connected in series with two 4 Ω resistors, which are connected in parallel, across a 30 V supply. The power dissipated in each 4 Ω resistor is 16 W. Calculate the value of the unknown resistor.

27 Resistors of 5 Ω, 10 Ω and 30 Ω are connected in parallel. The combination is connected in series with a 7 Ω resistor across a 50 V supply. Calculate the power dissipated in each resistor.

28 A 60 Ω resistor, and a resistor of unknown value, are connected in series with each other and with a parallel combination of a 200 Ω and a 50 Ω resistor. This circuit is connected to a 240 V supply. If the power dissipated in the circuit is 480 W, what is the value of the unknown resistor?

29 An electric radiator has a resistance of 19 Ω and is connected to a supply by a cable of resistance 1 Ω. A second radiator of 38 Ω resistance is connected to the supply by a cable of resistance 2 Ω. The supply voltage is 240 V. Calculate
    (a) the current in each radiator,
    (b) the total power taken from the mains. (C & G)

30 A domestic load comprises twelve 60 W lamps, eight 100 W lamps, two 2 kW radiators and a 3 kW immersion heater. The lamps are fed from a lighting circuit and the power from a ring-main circuit. The supply is at 240 V. What is the current in each circuit? Suggest a suitable fuse for each circuit from ratings of 2 A, 5 A, 10 A, 13 A, 20 A and 30 A. (C & G)

31 IEE Wiring Regulation 553-14 indicates that the total rating of a lighting circuit must not exceed 16 A. What is the maximum number of 100 W lamps wired on such a circuit if the supply voltage is 240 V? (*Note:* It would be very bad practice under normal circumstances to wire as many lights as this on one circuit.)

32 A small flat has the following loads:
    3 × 100 W lamps in use 5 h each day
    1 × 2 kW heater in use 4 h each day
    1 × 3 kW immersion heater in use 3 h each day
    1 × 4 kW cooker in use 2 h each day.
    The tariff for the supply of electricity is a fixed charge of 18p per week, plus an energy charge of 5p per unit. Find the total electricity cost for one week.

33 A water heater is 83·7% efficient, and is required to heat 12 l of water from 15°C to 75°C in 1 h. What minimum rating is required for the element? The specific heat of water is 4187 J/kg K.

34 How long will a 3 kW heater take to raise the temperature of 30 l of water from 20°C to 90°C if 10% of the heat supplied is lost in the process? The specific heat of water is 4187 J/kg K.

35 A builder's hoist lifts 100 kg of bricks through 40 m in 20 s. If the hoist is 75% efficient, what is the rating of the electric motor driving it?

36 If the motor of the previous exercise is fed from a 200 V d.c. supply and is 80% efficient, calculate its running current.

37 A lift motor takes a current of 20 A from a 300 V d.c. supply. The lift gearing is 50% efficient. If the lift is able to raise a load of 1440 N through a height of 50 m in 30 s, what is the efficiency of the motor?

38 Calculate the output of a motor with an input of 24 kW and an efficiency of 90%.

39 A motor with an output of 10 kW and an efficiency of 85% drives a drilling machine which is 60% efficient. What is the output of the machine, and what is the input of the motor?

Chapter 6

Permanent magnetism and electromagnetism

6.1 MAGNETIC FIELDS

Every schoolboy is familiar with permanent magnets, which can be used to pick up small iron objects. Such a magnet gives rise to a magnetic field which occupies the space in which the effects of the magnet can be detected. The magnetic field extends outwards in the space surrounding the magnet, getting weaker as the distance from the magnet increases. There are several methods of detecting the presence of a magnetic field, although it is quite invisible and does not affect other human senses. Theoretically, the magnetic field due to a magnet extends for considerable distances, but, in practice, it will combine with the fields of other magnets to form a composite field, so that the effects of most magnets can only be detected quite close to them.

Because we cannot see, feel, smell, or hear it, a magnetic field is difficult to represent. A 'picture' of a magnetic field is often useful, however, in deciding what its effects will be and to do this we consider that a magnetic field is made up of imaginary **lines of magnetic flux** which have the following properties:

(a) They always form complete, closed loops.
(b) They never cross one another.
(c) They have a definite direction.
(d) They try to contract as if they were stretched elastic threads.
(e) They repel one another when lying side by side and having the same direction.

It is most important to remember that any consideration involving lines of magnetic flux is quite imaginary. **The lines of magnetic flux do not exist, but to pretend that they do gives us a method of understanding the behaviour of a magnetic field.**

Experiments with crude permanent magnets were carried out many thousands of years ago by the ancient Greeks. They found that, if a magnet is suspended so that it can pivot freely, it will come to rest with a particular part (one end, if it is a bar magnet) pointing North. This is the **North-seeking pole** or **north pole** of the magnet, and the other end, which points South, is the **South-seeking pole** or **south pole**. If two magnets, with their poles marked, are brought together, the basic laws of magnetic attraction and repulsion can easily be demonstrated. These are:

(i) *Like poles repel.* For instance, two north poles or two south poles will try to push apart from each other.
(ii) *Unlike poles attract.* For instance, a north pole and a south pole will attract one another.

The Earth behaves as if a huge bar magnet were inside it, the south pole of the magnet being near the geographic North Pole, and the north pole at the geographic South Pole. The positions of the magnetic north and south poles of the Earth vary slowly with time, for reasons not yet fully explained. As far as we can tell, however, the Earth is a truly permanent magnet. If a piece of soft iron is magnetised by rubbing it in one direction with one pole of a permanent magnet, the soft iron will lose its magnetism slowly as time goes by. The magnetism could quickly be removed by heating or hammering the soft iron. Magnets which do not deteriorate with time, and which resist demagnetisation by ill treatment, are now available, and are mentioned later in this chapter.

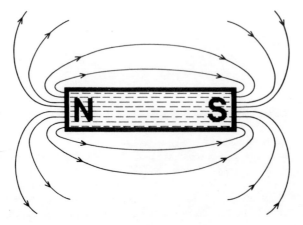

Fig. 6.1 Magnetic field due to bar magnet

The pattern of the lines of magnetic flux around a permanent bar magnet are shown in Fig. 6.1. This shows that the direction of the lines of magnetic flux (property c) is from north pole to south pole outside the magnet. These lines of magnetic flux obey the five rules given, their curved shapes outside the magnet being a compromise between properties d and e.

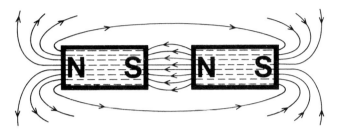

Fig. 6.2 Magnetic field due to two bar magnets, with unlike poles adjacent

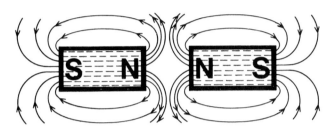

Fig. 6.3 Magnetic field due to two bar magnets with like poles adjacent

Fig. 6.2 shows the flux pattern due to two magnets with unlike poles close together. Lines of magnetic flux are imagined to try to contract (property d), and the magnets try to pull together.

Fig. 6.3 shows the magnetic-flux pattern due to two magnets with like poles close together. Since magnetic flux lines running side by side with the same direction repel, the two poles try to push apart.

These explanations indicate the neat way in which the results of magnetic fields can be forecast by the use of lines of magnetic flux. These lines can be plotted for an actual magnetic field in two ways. The first is to cover the magnet system concerned with a sheet of paper and then to sprinkle iron filings, which set themselves in the magnetic-flux pattern. The second method involves the use of a plotting compass, which consists of a miniature magnetic needle pivoted within a clear container. The compasss is placed on the paper, and the position of its north and south poles marked with a pencil. It is then moved, and its position adjusted until its south pole is at the mark indicating the previous position of the north pole. The new position of the north pole is marked, and the process repeated as often as is necessary. The line of marks which results is joined to form a line of magnetic flux. As many lines as are required can be produced in a similar way.

6.2 UNITS OF MAGNETIC FLUX

If the imaginary lines of magnetic flux already introduced actually existed, they could be counted and used as a measure of the quantity of magnetism. They do not exist, but the idea of a larger number of lines representing a greating magnetic flux than a smaller number is a useful one.

The general symbol for magnetic flux is Φ (Greek letter, capital 'phi'). The unit of magnetic flux is the **weber** (pronounced 'vayber' and abbreviated to Wb). It must be clearly understood that the weber is a measure of a total amount, or quantity, of magnetic flux, and not a measure of its concentration or density. Flux density is very important in some machines, and will depend on the amount of magnetic flux (the number of lines, in effect) which is concentrated in a given cross-sectional area of the flux path. The strength of a magnetic field is measured in terms of its **flux density** (symbol B), measured in **webers per square metre** (Wb/m^2), also called **teslas** (T). Thus, one weber of magnetic flux spread evenly throughout a cross-sectional area of one square metre results in a flux density of one tesla, or one weber per square metre. Similarly 1 Wb spread over $10\,m^2$ will give a flux density of $0\cdot 1\,T$.

80 Permanent magnetism and electromagnetism

Thus
$$B = \frac{\Phi}{A}$$

Where
B = magnetic-flux density, T (Wb/m^2)
Φ = magnetic flux, Wb
A = cross-sectional area of flux path, m^2

It should be appreciated that magnetic-flux path areas of the order of square metres are only met in the largest machines, but small quantities of flux in small areas can give rise to high flux densities.

EXAMPLE 6.1

The magnetic flux per pole in a d.c. machine is 2 mWb, and the effective poleface dimensions are 0·1 m × 0·2 m. Find the average flux density at the poleface.

Since the same submultiple prefixes apply to magnetic flux as to other quantities, 2 mWb = 0·002 Wb.

$$\text{Effective poleface area} = 0\cdot1 \times 0\cdot2 \text{ square metres}$$
$$= 0\cdot02 \text{ m}^2$$
$$B = \frac{\Phi}{A}$$
$$= \frac{0\cdot002}{0\cdot02} \text{ tesla}$$
$$= 0\cdot1 \text{ T}$$

EXAMPLE 6.2

A lifting electromagnet has a working flux density of 1 T, and the effective area of one poleface is a circle of diameter 70 mm. What is the total magnetic flux produced?

$$\text{Effective poleface area} = \frac{\pi d^2}{4}$$
$$= \frac{22}{7} \times \frac{70 \times 70}{4} \text{ square millimetres}$$
$$= 3850 \text{ mm}^2$$
$$= \frac{3850}{1000 \times 1000} \text{ square metres}$$
$$= 0\cdot00385 \text{ m}^2$$
$$\Phi = BA$$
$$= 1 \times 0\cdot00385 \text{ webers}$$
$$= 0\cdot00385 \text{ Wb}$$
$$= 3\cdot85 \text{ mWb}$$

6.3 ELECTROMAGNET

If an electric conductor is arranged to pass vertically through a horizontal sheet of paper, iron filings or a plotting compass will show a field pattern at the paper if a sufficiently large current flows through the conductor. Since both filings and compass are only affected by a magnetic field, this indicates that the current in the conductor must be producing the field; this observation is borne out in practice, because the effect on both filings and compass disappears when the current is switched off.

The arrangement described is shown in Fig. 6.4. This only shows a small part of the magnetic field, however, which extends for the full length of the conductor in the form of concentric rings of magnetic flux centred on the conductor. It is often necessary to indicate the direction of current flow in a cross-section of a cable, since cable and resultant magnetic field form a 3-dimensional arrangement. The convention for such an indication is shown in Fig. 6.5, current flowing away from the viewer (into the paper) being shown by a cross, and current flowing towards the viewer (out of the paper) being shown as a dot. This can easily be remembered if the current is replaced by a dart sliding in a hollow tube which represents the conductor. When the dart is sliding away from the viewer, its flights are seen as a cross, and, when sliding towards the viewer, the point is seen as a dot.

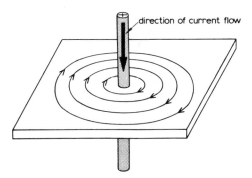

Fig. 6.4 Plotting magnetic field due to current-carrying conductor

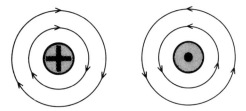

Fig. 6.5 Direction of magnetic field round current-carrying conductor

Using a plotting compass, the magnetic fields round current-carrying conductors can be shown to have the directions given in Fig. 6.5; that is, clockwise for current flowing away from, and anticlockwise for current flowing towards, the viewer. This can easily be remembered by the **screw rule**, which states that, if a normal right-hand thread screw is driven along the conductor in the direction taken by the current, its direction of rotation will be the direction of the magnetic field. It is quite common in electrical work for current-carrying conductors to lie side by side in a cable or conduit. Fig. 6.6 shows that the magnetic field in the space between the conductors may be quite intense if they carry current in opposite directions, although the magnetic field in an enclosing steel conduit will be in different directions in opposite sides of the conduit. If one or more conductors in a conduit carry current in the same direction, most of the magnetic flux will be restricted to the conduit, and may give rise to heating losses. This is the reason why feed and return cables should be enclosed in the same conduit.

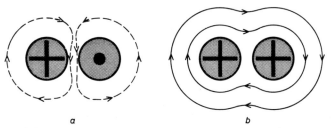

a *b*

Fig. 6.6 Magnetic fields due to two current-carrying conductors

The strength of the magnetic field round a conductor depends on the current, but, even at high currents, it is comparatively weak. To get a stronger field, the magnetic effects of a number of conductors can be added. The most common form for this arrangement is one long insulated conductor wound in a tight coil called a **solenoid**. The form taken by the solenoid is shown in Fig. 6.7*a*; Fig. 6.7*b* is a cross-section of the solenoid showing how the individual magnetic fields due to each separate turn merge to form a stronger field that is very similar to that of the permanent bar magnet (Fig. 6.1). The strength of the magnetic field produced depends on the current and on the number of turns used. Additional turns may be wound on top of the first layer to form a multilayer coil.

The polarity of the solenoid can be found easily by sketching it with correct current directions (Fig. 6.7b), or by using the **NS rule**. If arrows drawn on the ends of a capital N point in the direction of current flow in the solenoid when viewed from one end, that end is a north pole. If arrows on a capital S give the direction of current, the solenoid end viewed is a south pole (Fig. 6.8).

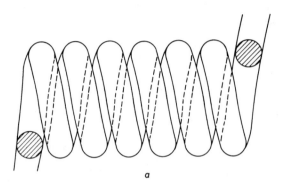

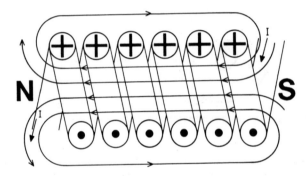

Fig. 6.7 Solenoid and its magnetic field

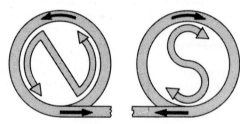

Fig. 6.8 NS rule

The solenoid is the electrical equivalent of the permanent magnet, but is often more versatile. For instance, the flux and flux density can be altered by a variation of solenoid current, or reduced to zero by switching off the current. Since the solenoid is simply a coil of conductor, its size and shape can be constructed to almost any requirements. The electromagnet, based on the solenoid, is the basis of many items of electrical equipment, of which the relay, the contactor, the motor, the generator, the transformer, and the telephone are examples.

6.4 CALCULATIONS FOR AIRCORED SOLENOIDS

If we multiply the number of turns of a solenoid by the current it carries, we arrive at the **magnetomotive force** (m.m.f.) of the solenoid, which is measured in **ampere turns** (At) or, more strictly, in **amperes** (A):

$$\text{Magnetomotive force} = \text{amperes} \times \text{turns}$$

For instance, two of the methods of producing an m.m.f. of, say, 1000 At, are 1 A flowing in a 1000-turn coil, or 10 A flowing in a 100-turn coil.

The magnetomotive force obviously affects the flux set up by the solenoid, but so, too, does the length of path taken by individual lines of magnetic flux. For instance, a 1000-turn coil wound in four layers and having a length of, say, 0·1 m will set up more flux for a given current than a 1000-turn coil wound in one layer over a length of 0·4 m. The path taken by the lines of magnetic flux will be longer in the second case, so that the magnetomotive force is 'stretched' over a greater distance. The magnetomotive force applied to each metre length of the path taken by the lines of magnetic flux is called the **magnetising force** (symbol H), measured in **ampere turns per metre** (At/m) or, more strictly, in amperes per metre (A/m)

$$H = \frac{\text{amperes} \times \text{turns}}{\text{length of flux path in metres}}$$

or

$$H = \frac{IN}{l} \text{ ampere turns per metre}$$

where

H = magnetising force, At/m or A/m
I = solenoid current, A
N = number of turns of conductor on solenoid
l = mean length of magnetic-flux path, m

The path taken by the lines of magnetic flux is often referred to as the **magnetic circuit**.

EXAMPLE 6.3

A solenoid, wound with wire having an overall diameter of 1 mm, must have 12 layers of winding, each 100 mm long. Calculate how much current must flow in the coil to give a magnetising force of 3000 At/m in a magnetic circuit of average length 0·25 m which has the coil mounted on it.

$$\text{Number of turns per layer} = \frac{\text{solenoid length}}{\text{wire diameter}} = \frac{100 \text{ mm}}{1 \text{ mm}} = 100 \text{ turns}$$

Total turns = 1200

$$H = \frac{IN}{l}$$

Therefore

$$I = \frac{Hl}{N}$$

$$= \frac{3000 \times 0.25}{1200} \text{ amperes}$$

$$= 0.625 \text{ A}$$

When a current flows in a coil, the resulting magnetising force sets up a magnetic flux. The amount of flux set up in air and other materials which are not attracted by a magnet will depend directly on the applied magnetising force. For instance, the extremely high magnetising force of one million ampere turns per metre would produce within the solenoid a flux density of $4\pi/10$ or 1·257 T. Thus the ratio

$$\frac{B}{H} = \frac{4\pi/10}{10^6} = \frac{4\pi}{10^7} \text{ or } 4\pi \times 10^{-7}$$

This constant relationship between flux density and magnetising force for air material is called the **permeability of free space**, and is given the symbol μ_0 (Greek letter 'mu').

$$\mu_0 = \frac{4\pi}{10^7}$$

EXAMPLE 6.4

An aircored solenoid is in the form of a closed ring (or toroid) of mean length 0·2 m and cross-sectional area 1000 mm². It is wound with 1000 turns and carries a current of 2 A. Find the magnetic-flux density and total magnetic flux produced within the solenoid.

$$H = \frac{IN}{l} = \frac{2 \times 1000}{0\cdot 2} \text{ ampere turns per metre} = 10\,000 \text{ At/m}$$

$$\frac{B}{H} = \mu_0$$

$$B = \mu_0 H$$

$$= \frac{4\pi}{10^7} \times 10\,000 \text{ tesla}$$

$$= \frac{4\pi}{1000} \text{ tesla}$$

$$= 0\cdot 01257 \text{ T}$$

$$\text{or } 12\cdot 57 \text{ mT}$$

$$\text{Total flux } \Phi = BA$$

$$= 0\cdot 01257 \times \frac{1000}{10^6} \text{ weber} \quad (\textit{Note: } 1\text{m}^2 = 10^6 \text{ mm}^2)$$

$$= 0\cdot 00001257 \text{ Wb}$$

$$\text{or } 12\cdot 57 \,\mu\text{Wb}$$

Notice the low magnetic-flux density and total flux set up by the comparatively high magnetising force.

EXAMPLE 6.5

A solenoid of 10 000 turns is wound on a brass ring of mean diameter 100 mm and cross-sectional area $0\cdot 01\,\text{m}^2$. How much current must flow in the solenoid to produce a total flux of 5 mWb?

Brass is a nonmagnetic material, so the rules given will apply.

$$B = \frac{\Phi}{A}$$

$$= \frac{0\cdot 005}{0\cdot 01} \text{ tesla}$$

$$= 0\cdot 5 \text{ T}$$

$$\frac{B}{H} = \mu_0$$

so
$$H = \frac{B}{\mu_0}$$

$$= \frac{IN}{l}$$

$$H = \frac{0\cdot 5 \times 10^7}{4\pi} \text{ ampere turns per metre}$$

Also,
$$H = \frac{I \times 10\,000}{\pi/10} \text{ ampere turns per metre}$$

$$I = \frac{0\cdot 5 \times 10^7 \times \pi}{4\pi \times 10\,000 \times 10} \text{ amperes}$$

$$= 12\cdot 5 \text{ A}$$

6.5 EFFECT OF IRON ON MAGNETIC CIRCUIT

Most materials in use for general purposes are nonmagnetic; that is, they show no magnetic properties. A simple test for magnetic properties is to see if a material is attracted to a permanent magnet, or to an electromagnet. The materials which show this attraction are iron, nickel and cobalt, as well as alloys containing one or all of them.

If a core of magnetic material is slid into a solenoid, the magnetic flux produced for the same current will be increased very many times. The ratio of flux produced by a solenoid with a magnetic core to flux produced by the same solenoid with an aircore (the current being the same in both cases) is called the **relative permeability** (symbol μ_r) of the material under these conditions. The value of the relative permeability of a nonmagnetic material is unity (one), whereas values for magnetic materials vary between wide limits, typical figures for common magnetic materials being from about 150 to about 1200. This value is not a constant for a given material, since it depends on the magnetising force applied. Taking relative permeability into account,

$$\frac{B}{H} = \mu_0 \mu_r$$

The product $\mu_0 \mu_r$ is called the **absolute permeability** of the material under the given conditions.

In many applications, a magnetic circuit has a small airgap in it. In some, such as moving-coil instruments or machines, this is because free movement must be possible between different parts of the magnetic circuit. In other machines, poorly fitting joints in the magnetic circuit have the same effect as an airgap. Even a small airgap may have a disproportionate effect on a magnetic circuit. Example 6.6 illustrates the advantage of using a magnetic material for a magnetic circuit, and the reduction of flux that occurs if an airgap is introduced.

EXAMPLE 6.6

(a) A solenoid is made to be identical with that described in example 6.4, but is wound on a wrought-iron core. When the solenoid current is 0·2 A, the relative permeability of the wrought iron is 500. Calculate the flux density and total flux in the ring.

(b) What would be the effect on flux and flux density if a radial sawcut were to be made in the wrought-iron core without disturbing the magnetising coil?

(a)
$$H = \frac{IN}{l}$$
$$= \frac{0\cdot 2 \times 1000}{0\cdot 2} \text{ ampere turns per metre}$$
$$= 1000 \text{ At/m}$$

$$\frac{B}{H} = \mu_0 \mu_r$$

so $B = \mu_0 \mu_r H$

Therefore
$$B = \frac{4\pi}{10^7} \times 500 \times 1000 \text{ tesla}$$
$$= 0\cdot 629 \text{ T}$$

$$\Phi = BA$$
$$= 0\cdot 629 \times 0\cdot 001 \text{ weber}$$
$$= 0\cdot 000629 \text{ weber}$$
$$= 0\cdot 629 \text{ mWb}$$

(b) The introduction of an airgap into the wrought-iron magnetic circuit will reduce considerably the magnetic flux and flux density set up by the magnetising coil. Calculations of flux values in a case of this sort are complicated, but it can be shown that, if the sawcut is only 3 mm wide, flux density will be reduced to 0·0714 T and flux to 0·0714 mWb.

The results of this example are worthy of close examination, since they show some of the principles of magnetic-circuit design. Part *a* shows that the substitution of a magnetic material as the core of the solenoid has increased the flux and flux density by 50 times, even though the magnetising current is one-tenth its previous value.

The answer to part b shows that the airgap (length 3 mm) makes the establishment of magnetic flux nearly ten times as difficult as when the iron path was complete. This is despite the fact that 98·5% of the circuit (197 mm) still consists of iron.

6.6 PERMANENT MAGNETS

Permanent-magnet materials have been the subject of intense development over the past few years, and magnets can now be made that really are permanent, despite the most adverse conditions of temperature, mechanical ill treatment and demagnetising fields. The materials used are the three main magnetic elements (iron, nickel and cobalt), with the addition of other elements such as tungsten, carbon, manganese, chromium, copper, aluminium etc. Modern permanent magnets are often impossible to drill or machine, and must be cast in their final correct shapes. Intricate magnetic circuits often use a small permanent magnet at the centre of soft-iron extension pieces.

Although modern permanent magnets are unlikely to change their properties during normal use, they can sometimes be damaged magnetically when removed from their surrounding magnetic circuit. If it is necessary to remove a permanent magnet, the following precautions should be observed:

(a) Complete the magnetic circuit using a mild-steel 'keeper'.
(b) Do not slide the keeper against the magnet; remove it with a direct pull.
(c) Do not touch the magnet with steel tools, such as screwdrivers or spanners.
(d) Do not put the magnet on a steel-topped workbench.
(e) Keep the magnet in a nonmagnetic tray or box, well separated from other magnets.
(f) Do not allow people to play with the magnet.

6.7 SUMMARY OF FORMULAS FOR CHAPTER 6

$$B = \frac{\Phi}{A} \qquad \Phi = BA \qquad A = \frac{\Phi}{B}$$

where

B = magnetic-flux density, T (Wb/m^2)
Φ = magnetic flux, Wb
A = cross-sectional area of flux path, m^2

$$H = \frac{IN}{l} \qquad I = \frac{Hl}{N} \qquad N = \frac{Hl}{I}$$

where

H = magnetising force (ampere turns per metre, or amperes per metre)
I = magnetising current, A
N = number of magnetising coil turns
l = mean length of magnetic-flux path, m

Permeability of free space, $\mu_0 = \dfrac{4\pi}{10^7}$

$$\mu_0 = \frac{B}{H} \qquad B = \mu_0 H \qquad H = \frac{B}{\mu_0}$$

For magnetic materials,

$$\mu_0 \mu_r = \frac{B}{H}, \qquad B = \mu_0 \mu_r H, \qquad H = \frac{B}{\mu_0 \mu_r}$$

where μ_r is the relative permeability of the magnetic material under the given conditions.

6.8 EXERCISES

1   It is required to determine the magnetic field surrounding a straight bar magnet. Give a brief description of a laboratory procedure necessary to produce this. (NCTEC)

2   Draw a diagram to show a solenoid wound over an iron core. Mark on the diagram the magnetic lines of flux resulting from current in the solenoid. Show a direction for the current, and give the resulting direction of the flux lines.

3 Draw circles to represent the cross-sections of two conductors lying side by side. Mark the directions of currents in the conductors, and sketch the resulting magnetic field if the two conductors carry currents
   (a) in the same direction,
   (b) in opposite directions.

4 The total flux required in the core of a power transformer is 0·156 Wb. If the area of the core is 0·12 m², what will be the flux density?

5 A moving-coil instrument has an airgap of effective cross-sectional area 56 mm². What is the total flux if the gap flux density is 0·12 T?

6 A coil of 400 turns wound over a wooden ring of mean circumference 0·5 m and uniform cross-sectional area 800 mm² carries a current of 6 A. Calculate
   (a) magnetising force,
   (b) total flux,
   (c) flux density.

7 If an airgap of length 20 mm is introduced into a magnetic circuit of cross-sectional area 0·009 m² that is carrying a flux of 1·2 mWb, what extra magnetomotive force will be required to maintain the flux?

8 Calculate the flux and flux density set up by the solenoid of exercise 6 if the wooden ring is replaced by one of magnetic material having identical dimensions and a relative permeability of 200 at this flux density.

9 A coil of insulated wire of 500 turns and of resistance 4 Ω is closely wound on an iron ring. The ring has a mean diameter of 0·25 m and a uniform cross-sectional area of 700 mm².
   Calculate the total flux in the ring when a d.c. supply at 6 V is applied to the ends of the winding. Assume a relative permeability of 550.
   Explain the general effect of making a small airgap by cutting the iron ring radially at one point. (C & G)

10 Make sketches indicating current direction and resulting magnetic field when current flows in (a) a straight conductor, (b) a solenoid about 40 mm in diameter and 50 mm long, (c) a similar solenoid with a 20 mm diameter steel core 50 mm long, (d) a solenoid with a steel core in the shape of a closed ring. (C & G)

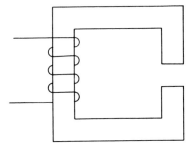

Fig. 6.9 Diagram for exercise 11

11 Fig. 6.9 shows a coil of wire wound on to an iron core with an airgap in one limb. How would the magnetic flux in the gap be affected by changes in (a) the current in the coil, (b) the number of turns of wire, (c) the length of the iron path, (d) the length of the gap, (e) the permeability of the iron? (C & G)

Chapter 7
Applications of electromagnetism

7.1 INTRODUCTION

This chapter will describe some of the applications of electromagnetism which are commonly in use. It should be noted, however, that no complete list of such applications is possible, because this list is almost endless. The principles put forward in Chapter 6 apply to all these devices.

Two of the most important of all the electromagnetic equipments are not mentioned in this chapter. These are the generator and the motor, which are introduced in Chapters 9 and 11, respectively.

7.2 BELLS AND BUZZERS

The operation of bells and buzzers is identical, and units of these types are used widely to give audible warnings. There are a number of types of bell.

The **single-stroke bell** has applications where simple signalling is necessary. It consists of an electromagnet, which, when energised, attracts a soft-iron strip, called an armature, against the pull of a flat spring strip.

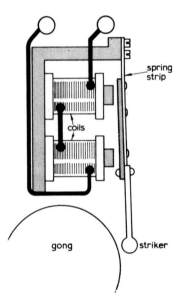

Fig. 7.1 Single-stroke bell

A striker supported by the armature sounds the gong (Fig. 7.1). When the current is switched off by releasing the signalling push, the spring strip returns the armature to its original position.

The **trembler bell** is the most widely used type, and is similar in construction to the single-stroke bell, but has the addition of a set of contacts, P, which are opened by movement of the armature towards the electromagnet (Fig. 7.2). This de-energises the magnet, and allows the armature to return to its original position, closing the contacts and repeating the cycle. While the supply is maintained, the armature is in a continual state of vibration, its striker hitting the gong repeatedly. The rate of striking depends on the flexibility of the spring strip, and the weight and size of the moving system.

The **buzzer** has no striker and normally oscillates more rapidly; some adjustment of movement, and hence of buzzer note, being possible with the contact screw.

The **continuous ringing bell** is particularly useful for alarm systems, since it will continue to ring after the external operating circuit has been broken. The construction (Fig. 7.3) includes a relay arm which drops when the bell first rings, closing an internal operating circuit, which continues to ring the bell until reset by lifting the relay arm to its original position. This can be done by means of a pull cord or by a solenoid, operated remotely.

The **polarised bell** is mainly used for telephone circuits where a low-frequency a.c. supply is used. A permanent-magnet system (Fig. 7.4) is fitted with two operating coils connected in series but wound in opposite directions. Alternating current will alternately weaken and strengthen the magnetic field in the side limbs, one being strengthened while the other is weakened. The moving system will thus be attracted to each side in turn, the striker sounding each bell in turn.

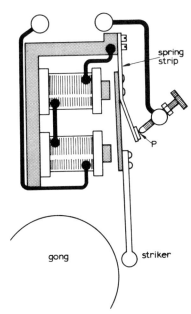

Fig. 7.2 Trembler bell

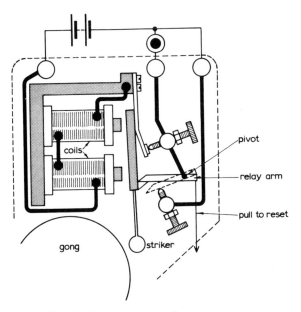

Fig. 7.3 Continuous-ringing bell

7.3 BELL INDICATORS AND CIRCUITS

There are many applications of systems that require that a large number of widely spaced pushes should be used to indicate, at one central position, that service is required. Hospitals, hotels, restaurants and many other semipublic buildings need to be so equipped. Fire-alarm and burglar-alarm systems are other examples of installations where the point of origin of the alarm may need to be indicated at some central or remote point.

The range of bells and buzzers with differing tones is strictly limited, so the indicator board was developed to cater for this situation. Usually, only one bell or buzzer sounds, the board giving a clear indication of the position of the push that was operated.

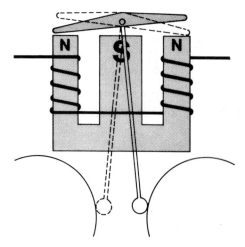

Fig. 7.4 Polarised bell

There are three basic types of indicator unit:

(a) **Pendulum-type indicator:** The indicator 'flag' is attracted to a solenoid while the press is pushed, falling away and swinging to give indication when it is released. This is a simple and inexpensive system, but the indication ceases after a time when the flag stops swinging. The pendulum type is now being replaced by the other two types.

(b) **Mechanical-reset-type indicator:** The coloured disc of this unit is not normally seen through the window on the front of the indicator, but is pulled into view by a solenoid coil when the circuit concerned is energised. It then remains in the indicating position until reset by the mechanical operation of a rod or lever.

(c) **Electrical-reset-type indicator:** This is similar in operation to the mechanical-reset type, but is returned to the normal position by a reset solenoid operated from a control push normally situated adjacent to the board (Fig. 7.5).

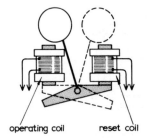

Fig. 7.5 Electrical reset-type bell indicator unit

Units of these types are made up into boards containing any required number of ways, each way suitably labelled.

Fig. 7.6 shows how a four-way indicator board can be connected. Additional boards may be connected in series with the first if indication at other points is necessary.

7.4 RELAYS AND CONTACTORS

Relays

A relay is similar in construction to the single-stroke bell, but a set of contacts are opened or closed by operation, instead of a gong being struck. This enables one circuit to operate another (Fig. 7.7), a very low current often being sufficient to close the contacts of a relay, and thus operate a high-power circuit. Some relays have a large number of contacts, and can be used in complicated circuits for a wide variety of switching purposes. Fig. 7.8 shows a typical relay used by British Telecom for telephone systems.

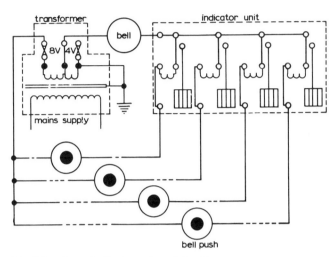

Fig. 7.6 4-way indicator unit and circuit

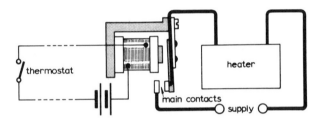

Fig. 7.7 Principle of relay and of contactor

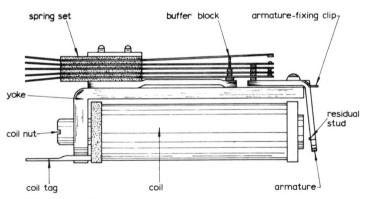

Fig. 7.8 Typical telephone relay

A relay usually has an operating coil wound on a magnetic circuit with a moving section, or armature, held open by a spring. When the operating current is switched on and sets up a magnetic field, the armature is closed and operates the contacts.

Contactors

Contactors are simply very large relays, enabling a heavy load to be switched on and off with a very small operating current. A few applications are

(a) motor starting: the contactor coil, and hence the motor, is controlled by 'stop' and 'start' pushbuttons; the contactor also ensures that the supply leads to the motor are broken in the event of a supply failure, so that the motor cannot restart automatically when the supply is restored

94 *Applications of electromagnetism*

(b) timeswitch contactor: a heavy heating load can be switched by a timeswitch with low-rated contacts; the timeswitch controls the coil of the contactor, which controls the load
(c) remote switching: the principle is shown in Fig. 7.7.

7.5 TELEPHONES

Operation of a telephone is based on the fact that noise consists of a series of pressure waves in the air. These pressure changes are caused by the vocal chords in speech, and alternate bands of high and low pressure move outwards in ever widening circles from the noise source, like the ripples on a pond into which a stone has been thrown. These waves vibrate the eardrums, making it possible to hear the sound.

The telephone changes these variations in pressure into corresponding changes in an electric current, which can be made to flow for long distances before giving up energy to convert the current variations back into pressure changes.

Microphone or transmitter

This is the device used to change sound into an electric current. Its operation depends on the fact that a mass of small **carbon granules** has electrical resistance which varies with the sound pressure applied to it.

The construction, shown in simplified form in Fig. 7.9, consists of two flat discs of solid carbon, one fixed solidly at the back of the microphone capsule and one fixed to the back of a thin metal diaphragm (usually of aluminium). The diaphragm vibrates in sympathy with the pressure waves striking it, and thus the carbon

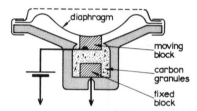

Fig. 7.9 Simplified diagram of carbon-granule microphone capsule

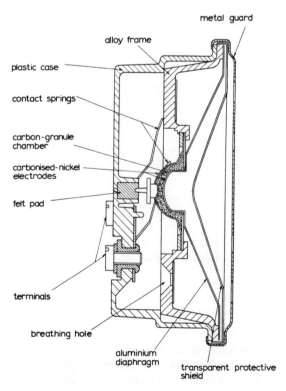

Fig. 7.10 Typical telephone transmitter inset

granules packed between the two carbon discs are subject to pressure variations. If a battery is connected as shown, the current will vary with the resistance of the granule pack, its waveshape being very similar to that of the sound concerned. For ease of servicing, microphone capsules are normally self-contained units which plug into the telephone handset. Fig. 7.10 shows in diagrammatic form the type of transmitter used in most public telephones.

Receiver

This is the device which converts the varying electric current back into sound waves, and consists basically of a permanent magnet wound with coils of fine wire (Fig. 7.11). A thin diaphragm of magnetic material is held in position by the permanent magnet, very close to its poles but not touching them. When the current

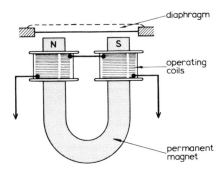

Fig. 7.11 Simplified diagram of permanent-magnet receiver

from the microphone passes through the coils, the magnetic field varies in sympathy with it, so that the diaphragm is vibrated. This vibration sets up pressure waves in the air, corresponding closely to those at the microphone.

Circuit

If two microphones and two receivers were connected in series with a battery (Fig. 7.12), a conversation could be held. However, the useful distance over which the system would operate satisfactorily would be

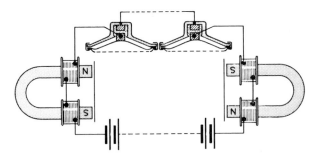

Fig. 7.12 Elementary telephone circuit

very limited. There is no ringing system, and the batteries would soon be exhausted, since they are in circuit all the time. Some simple circuits which overcome these difficulties are given in Section 7.6.

7.6 SIMPLE TELEPHONE CIRCUITS

A typical simple circuit for two telephones is shown in Fig. 7.13. If the press key of one telephone is operated, the circuit is completed as shown by the open arrows, the battery associated with one telephone ringing the bell of the other. It should be noted that both handsets must be in position for the ringing circuit to be complete, since lifting either set results in a break in the circuit. When both sets are lifted, however, the speaking circuit is completed as shown by the closed arrows.

The power loss in a resistive line is proportional to the square of the current flowing, so that, if the current can be reduced ten times, the loss will be reduced 100 times. A longer circuit is possible here, the reduction of current being obtained by using induction coils. A complete circuit of this type is shown in Fig. 7.14, from which it should be noted that the receiver is connected in the low-current system, where it operates quite satisfactorily. The induction coils are, in fact, simple transformers, which convert the high varying direct current of the microphone circuit to a low-current, higher-voltage system of the same waveshape in the remainder of the circuit.

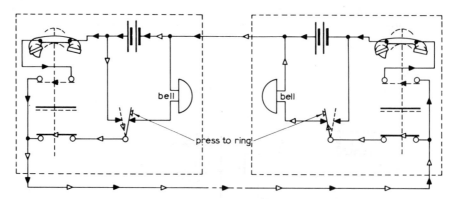

Fig. 7.13 Simple 2-instrument telephone circuit

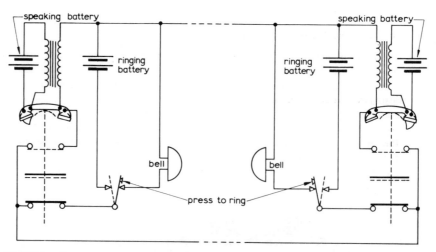

Fig. 7.14 Simple telephone circuit with induction coils

It must be stressed that the circuits given have been considerably simplified to show the basic principles concerned. Practical circuits can be very complex indeed, with large numbers of telephones in communication with each other and with so-called 'master' sets, which often have special facilities for priority and secret conversations.

The use of loudspeaker-type units is increasing, although its field is inclined to be limited by the publicity involved. A loudspeaker-type receiver and a particularly sensitive microphone take the place of the normal handset in such units.

7.7 LOUDSPEAKERS

The loudspeaker is used to convert electrical signals into sound waves. The very weak electrical pulses received by a radio aerial, for instance, are amplified in the electronic circuits of the radio set, before being fed to the loudspeaker so that the programme can be heard.

A typical loudspeaker has a magnetic circuit comprising a permanent magnet, and completed by soft-iron pole pieces, so that a strong magnetic flux exists in a short cylindrical airgap. A moving coil, called the speech or voice coil, is suspended from the end of a paper or plastic cone so that it lies in the gap (Fig. 7.15). Alternating currents in the coil push it up and down, vibrating the cone and producing sound waves.

The excellence of a loudspeaker depends on the strength of the magnetic field in the gap, the lightness of the coil, which is often wound of aluminium wire, and the size of the cone. A large cone produces low-frequency notes well, and a small cone deals best with high-frequency notes. Some loudspeakers have two cones to enable them to reproduce notes faithfully over a wide range of frequencies.

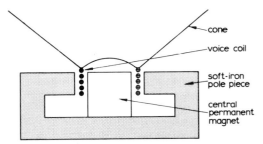

Fig. 7.15   Simple construction of permanent-magnet moving-coil loudspeaker

7.8 MOVING-IRON INSTRUMENTS

The moving-iron instrument is a device used for the measurement of electric current, which can also be adapted for use as a voltmeter. There are two basic types of moving-iron instrument.

Moving-iron repulsion instrument

In its simplest form, this instrument consists of a magnetising coil which carries the current to be measured, inside which are two parallel soft-iron bars (Fig. 7.16). One of the bars is fixed to the coil, and the other is attached to the moving system of the instrument and is free to swing with it. When current flows in the coil,

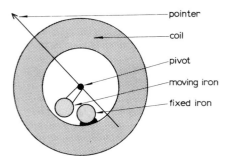

Fig. 7.16   Principle of moving-iron repulsion instrument

both bars are magnetised to an extent which depends on the current. Since the bars are magnetised by the same field, their magnetic polarities will be the same, with both north poles together at one end and both south poles together at the other. The repulsion which results moves the bar which is free to swing, and gives a deflection.

Moving-iron attraction instrument

The magnetising coil of this instrument is wound so that the central space will accept a soft-iron vane which is attracted to it (Fig. 7.17). The vane is attached to the moving system, and thus indicates the strength of the magnetic field, and hence of the current producing it.

All moving-iron instruments have the major advantage of reading on both d.c. and a.c. supplies. A coil fed with an alternating current results in an alternating magnetic field, which continues to attract the soft-iron vane, or to reverse the direction of the magnetisation of the two soft-iron bars together, so that they continue to repel. These instruments are robust and comparatively cheap, but have a serious disadvantage, in that their scales follow a 'square law' and are very cramped at the lower end. Little can be done to improve the lower end of the scale, but the rest of it can be made more linear by shaping the disc of the attraction type, or substituting shaped plates for soft-iron bars in the repulsion type. Other disadvantages are that moving-iron instruments are not as accurate as permanent-magnet moving-coil types (Section 11.6), and are affected by stray magnetic fields and by changes in the frequency of an a.c. supply.

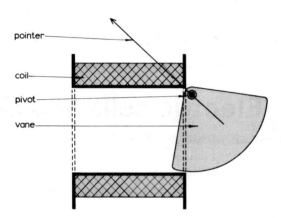

Fig. 7.17 Principle of moving-iron attraction instrument

7.9 EXERCISES

1. Describe the construction and operation of
   (a) an electric bell, and
   (b) an electric buzzer.

2. Describe the construction and operation of a single-stroke bell; indicate one application of such a bell.

3. Using a carefully drawn sketch, show the construction of a continuous-ringing bell. Explain how it functions, and where it could be used.

4. Describe one type of bell indicator. Draw a circuit diagram for a 4-way indicator system, using one bell, an indicator board, and four pushes.

5. Explain the construction and operation of a relay. How can this device allow heavy currents to be switched remotely, using small conductors to the remote position?

6. Suggest four applications for a relay or contactor, and explain each.

7. Describe, with the aid of a sketch, the construction and operation of a simple electromagnetic relay suitable for a bell circuit. (ULCI)

8. Make a neat, labelled sketch to show the construction and operation of *one* of the following:
   (a) a telephone receiver (earpiece)
   (b) a buzzer. (C & G)

9. With the aid of a sketch, give a brief description of the construction of a telephone receiver. (NCTEC)

10. Describe with the aid of a sketch the construction and operation of a telephone microphone.

11. Draw a circuit showing two telephones connected together. Each telephone must have provision for calling the other, as well as the speaking circuit. Explain how the circuit functions.

12. Using a clearly labelled diagram, describe the construction and operation of a loudspeaker.

13. Two permanent-magnet moving-coil loudspeakers are connected by a pair of wires to form a simple telephone system. Explain briefly how each can operate as both transmitter and receiver. (C & G)

14. Make a labelled sketch of *one* of the following:
    (a) the movement of a moving-coil permanent-magnet loudspeaker driving a conical diaphragm
    (b) a d.c. contactor suitable for 100 A
    (c) a 'telephone-type' relay. (C & G)

15. Describe the construction and operation of one type of moving-iron instrument. What are its advantages and disadvantages when compared with the permanent-magnet moving-coil instrument?

Chapter 8
Electric cells and batteries

8.1 STORING ELECTRICITY

Most of the electrical power used is generated in rotating machines; a sufficient number of generators are necessary to provide the maximum load required, since power used at a particular time must be generated at that time. Many generators stand idle for long periods because they are needed only to meet the peak demands, which usually occur for only a few hours a day. Fewer generators would be necessary if electricity could be stored during the night for use during the day. One method of storage is to provide chemical energy which can be converted to electrical energy as required. Although the cost of such an operation on a national scale would be prohibitive, this method is very widely used. A unit for chemical to electrical energy conversion is called a **cell**, and there are two types of cell.

Primary cells have the active chemicals placed in them when they are made. When the chemicals are used up, they are sometimes replaced, but it is more usual to throw away the spent cell. The majority of primary cells are now made in the dry form, and are widely used in torches, portable radio sets and the like.

Secondary cells are capable of being reactivated when their energy is spent. This is done by passing a charging current of electricity through them. Some of this energy is converted to the chemical form, and stored for future use. Secondary cells and batteries are widely used on vehicles, for standby supplies in case of mains failure, and similar purposes.

8.2 PRIMARY CELLS *once off use & is then discarded*

Primary cells are those which have their energy added in chemical form during manufacture, and which normally cannot be recharged once this energy is spent. The easiest way to understand primary cells is to first take the simplest of them and to examine its defects, showing how these are overcome in more complex cells.

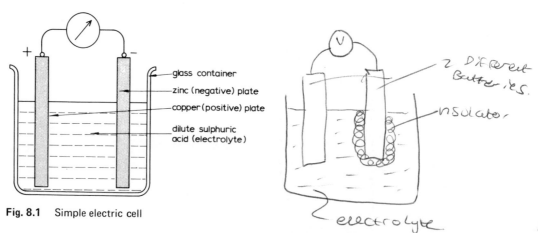

Fig. 8.1 Simple electric cell

Simple cell

Fig. 8.1 shows a simple cell, which consists of plates of copper and zinc immersed in a weak solution of sulphuric acid. If an external circuit is connected across the plates, current flows from the copper (positive) plate to the zinc (negative) plate, the circuit being completed through the sulphuric acid, which is called the **electrolyte**. Hydrogen is produced in this action and collects on the copper plate in the form of fine bubbles, which effectively insulate the plate from the electrolyte. This defect is called **polarisation**, and it results in a sharp decrease in cell e.m.f. from its initial value of 1·08 V, when a current is drawn.

The zinc plate is often in a form which is far from pure; small particles of other metals, such as iron and lead, are embedded in it. When the plate is immersed in the electrolyte, these impurities, in conjunction with the zinc, form tiny cells on its surface, and the zinc plate is eroded away. This defect is called **local action**. These two disadvantages make the simple cell unsuitable for practical use. Practical cells use pure-zinc plates, or a coating of mercury on the zinc, to prevent local action. Polarisation is overcome by placing the positive electrode in a chemical which absorbs the hydrogen, and is called a **depolariser**.

Wet Leclanché cell

Cells of this type are still in operation in some bell and telephone circuits. A section of a typical cell is shown in Fig. 8.2. The mercury-coated zinc (negative) rod is enclosed within the glass jar, in its electrolyte of ammonium chloride, sometimes called sal ammoniac. The carbon (positive) rod is packed in a depolariser of crushed carbon and manganese dioxide, separated from the electrolyte by a porous pot. The depolariser is efficient, but rather slow in action, so that the cell is most useful for intermittent operation. The electrolyte has a tendency to creep up the sides of the glass jar. This is prevented by painting or greasing the inside of the upper part of the jar. The e.m.f. of this cell is 1·5 V.

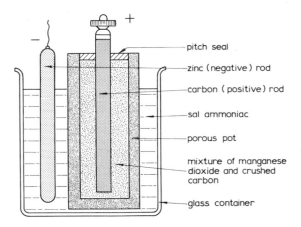

Fig. 8.2 Leclanché cell (wet)

Dry cell

This is a form of Leclanché cell in which the liquid is replaced by a paste, and which can thus be used in any position. The sheet-zinc container, often slid into a cardboard tube, holds the electrolyte, which consists of plaster of paris saturated with ammonium chloride (Fig. 8.3). Since the electrolyte is in paste form, it does not mix with the crushed carbon and manganese-dioxide depolariser, from which it is separated by

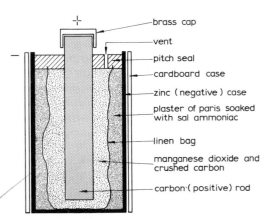

Fig. 8.3 Leclanché cell (dry)

enclosing the latter in a linen bag. At the centre of the bag is the carbon (positive) rod, which is usually surmounted by a brass cap to make easy contact. The cell is sealed with pitch, a vent being left for the escape of gases. The cell e.m.f. is 1·5 V when new, falling to a steady 1·4 V in service. When the cell voltage falls to 1 V, it should be discarded, since, at this stage, the zinc container quickly corrodes and allows its contents to escape. **Leakproof** cells enclosed in a second steel case are available to prevent this nuisance.

Mercury cells

These cells are the result of considerable research, and have such useful characteristics that the application is certain to increase. The basic form of the mercury flat cell is shown in Fig. 8.4, although other types of construction are used. The negative electrode is of amalgamated zinc, either as a foil or as a pressed powder, and the positive electrode is formed from graphite and mercuric oxide, which acts as a depolariser. The small percentage of powdered graphite is added to the mercuric oxide to reduce the internal resistance of the cell. The electrolyte, which is contained in absorbent material, consists of a solution of potassium hydroxide and dissolved zinc oxide.

Mercury cells have an e.m.f. of 1·35 V, and, although more expensive than most other forms of primary cell, they have many advantages. These include very high capacity, long shelf life without deterioration, low internal resistance, good voltage stability over a wide range of temperatures, leakproof construction, and the ability to withstand mechanical shock. Very small mercury cells are possible, and are widely used in hearing aids, watches and so on. Their reliability makes these cells an obvious choice for use in equipment where premature failure cannot be allowed.

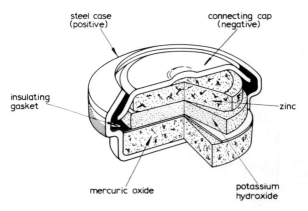

Fig. 8.4 Basic mercury flat cell

8.3 SECONDARY CELLS

There are two basic types of secondary cell, the lead-acid type and the alkaline type. Both types have a reversible chemical action and can be recharged by passing an electric current through them.

The characteristics, maintenance and charging of secondary cells is covered in Section 8.4.

Lead-acid cells

Lead-acid cells are the most widely used type of secondary cell, and consist of two plates of lead in an electrolyte of dilute sulphuric acid. The positive plate is often in the form of a lead-antimony-alloy lattice into which active lead oxide is pressed. The negative plate is usually of pure lead. Water is produced on discharge, and lowers the **specific gravity** of the dilute sulphuric acid. Specific gravity is a method of indicating the strength of the electrolyte, and is the ratio of the mass of a given volume of the electrolyte to the mass of the same volume of water. Measurement of specific gravity using a **hydrometer** is a good method of determining the state of charge. If the cell is overdischarged, or allowed to stand for long periods in the discharged condition, the lead sulphate coating on the plates becomes hard and is difficult to remove by charging. In this condition, the efficiency of the cell is reduced, and it is said to be **sulphated**. A healthy lead-acid cell has an initial e.m.f. in the region of 2·2 V, which falls to 2 V when in use.

The construction of lead-acid cells varies with the duty they are required to perform. Cells for use in vehicles are usually made up into batteries, a number of cells being included in separate containers in the same moulding of hard rubber, celluloid or polystyrene. Terminals for each cell are brought through the top of the moulding, which includes removable vent plugs for inspection and topping up. To obtain maximum capacity for minimum volume, the plates are mounted close together but prevented from touching by separators of thin wood or porous plastic.

Stationary cells are permanently installed in buildings to provide power for emergency lighting, bells, clocks, telephones etc. Since light weight is not so important as in vehicle cells, glass jars are used as containers. The plates are suspended into the jars, with glass rods or tubes as separators, and with a glass sheet overall to prevent acid spray when the cell is gassing. Closed tops are also used. Leakage currents are limited by mounting the containers on glass or porcelain insulators. Fig. 8.5 shows a view of an enclosed cell with its plates removed to show the construction.

All lead-acid cells have space left below the plates for accumulation of active material forced out of the plates while in use, and which might otherwise 'short' out the cell. The specific gravity of a charged cell varies with its type, and is also slightly affected by temperature, but an average value is 1·250.

Alkaline cells

These cells are enclosed in steel cases, and use potassium hydrate as an electrolyte. The specific gravity of the electrolyte remains steady during discharge at about 1·20, but may fall with age, and should be replaced when it reaches 1·16. There are several types of cell using this electrolyte.

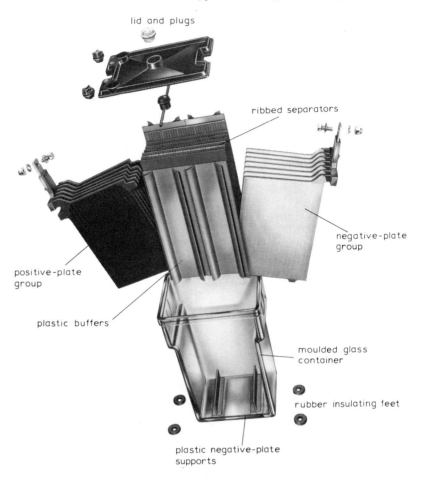

Fig. 8.5 Closed-top stationary cell dismantled to show construction

The **nickel-cadmium cell** has flat plates, interleaved as in the lead-acid cell. The positive plates are of nickel hydroxide, and the negative plates are of cadmium. Fig. 8.6 shows the construction of this type of cell.

Nickel-iron cells have positive plates made up of perforated tubes of nickel containing the active nickel hydroxide, the flat negative plates being of iron, but their manufacture is being discontinued.

The nickel-iron cell has the greater storage capacity for a given volume of the two types, has a longer life and can be subjected to severe overdischarge without detriment. Nickel-cadmium cells have lower internal resistance, and can thus provide heavy discharge currents for a low voltage drop. They maintain their charge for very long periods without attention. Both types are capable of operation at much higher temperatures than lead-acid cells.

The robustness of alkaline cells, and their resistance to both mechanical and electrical ill treatment, make them a suitable choice for some types of electric vehicle, for operation of high-voltage switchgear, for railway signalling, and for similar purposes where failures cannot be accepted. Their high cost, wide terminal-voltage range (e.m.f. varies from about 1·4 V charged to 1·1 V discharged), and higher internal resistance (more than double that of a lead-acid battery with the same terminal voltage), make them less popular for general use than lead-acid types.

8.4 CARE OF SECONDARY CELLS

When the chemicals of a cell have changed to the inactive form, they can be made active once more by passing a charging current through the battery in the opposite direction to the discharge current. The supply voltage must be in excess of the battery voltage, or no charging current can flow. There are two methods of charging:

(a) *Constant-voltage charging:* A constant voltage is applied to the battery under charge. While the battery is charging, a steady increase occurs in its terminal voltage, so that the charging current tapers off to a lower value at the end of the charge than at the beginning. This system has the advantage that no adjustments are required during charging; however, since it is recommended that lead-acid batteries are better charged at a constant rate, this method is seldom employed in the UK, although it is widely used on the Continent.

(b) *Constant-current charging:* This system uses either an adjustable voltage source or a variable resistance, so that the charging current can be kept constant throughout the charge. Healthier stationary batteries result from this method, but the variation concerned normally requires manual adjustment, although automatic means can be provided.

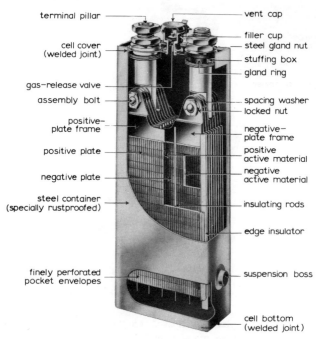

Fig. 8.6 Construction of nickel-cadmium cell

Capacity

The total charge which a battery or a cell will hold is measured in terms of the current which is supplied, multiplied by the time for which it flows. In practice, this figure varies depending on the rate of discharge, a quick discharge giving a lower figure than a slow one. Capacity (which should not be confused with capacitance) is usually measured at the ten-hour rate; for instance, a sixty ampere-hour (60 Ah) battery will provide six amperes for ten hours.

Charging and maintenance of lead-acid and alkaline cells are quite different, and will be considered separately.

Lead-acid cells: maintenance

These cells are the most widely used of the secondary cells, owing to their comparatively low cost and higher voltage per cell. They can be damaged, however, by charging or discharging too quickly, overcharging, leaving in the undischarged state etc. Healthy cells can only be maintained in condition, either by keeping them fully charged, or by periodically recharging, ideally at monthly intervals. A lead-acid cell will lose its charge if left standing over a period of a few months.

For periodic charging or for recharging after use, the constant-current method is generally applied, the current value needed varying somewhat with the type of cell. A common value is one-tenth of the ampere-hour capacity at the 10 h rate, i.e. 6 A for a 60 Ah battery. It is important not to overcharge. There are three methods by which the state of charge can be determined:

(i) *Colour of plates:* Fully charged cells should have clear light-grey negative plates, and rich chocolate brown positive plates.
(ii) *Terminal voltage:* Open-circuit voltage of a fully charged cell depends on the type, being from 2·1 V to 2·3 V. If this voltage is measured with the charging current flowing, it will be increased by the voltage drop in the internal resistance of the cell.
(iii) *Specific gravity:* This is the best method of determining the state of charge, the specific gravity of the electrolyte varying from 1·25 for a fully charged cell to about 1·17 for a discharged cell. These figures apply to storage batteries, and may vary slightly for other types of cell.

Specific gravity readings can be taken using a **hydrometer**, shown in Fig. 8.7. This consists of a wide glass tube with a rubber tube fitted to one end, and a rubber bulb to the other. The rubber tube is inserted into the electrolyte after squeezing the bulb, release of which will draw a small quantity of acid into the glass

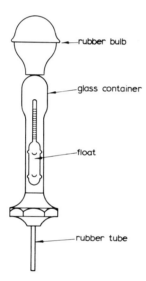

Fig. 8.7 Glass-bulb hydrometer

tube. This tube contains a small glass float. In a dense acid, the float is lifted high, but sinks lower in a weaker electrolyte: direct readings can be taken from graduations on the float neck.

Acid temperature will affect specific-gravity readings. The values given are for 15°C. For each degree Celsius above or below this figure, 0·0007 should be added or subtracted, respectively.

'Trickle' and 'float' charging
A lead-acid cell can be kept in a healthy condition for long periods by making good the losses in capacity as they occur. This is done by continually trickle-charging with a small current, the value of which in milliamperes is equal to the ampere-hour capacity for cells of up to 100 Ah (i.e. 60 mA for 60 Ah battery). Above this capacity, the correct trickle-charging rate is given by

$$(70 + 3 \times 10 \text{ hour capacity}) \text{ milliamperes}$$

Some d.c. systems use 'floating' batteries, of the same nominal voltage as the supply and connected directly across it. If the battery voltage falls, supply voltage is greater, and the battery charges until values are equal. If the supply fails, the battery at once takes over.

Precautions
Precautions must be carefully taken where lead-acid cells are used or charged, mainly because the cells 'gas', giving off hydrogen and oxygen during charge and discharge. The following are the most important points to remember:

(i) Keep the electrolyte at the correct level with distilled water, to make good the loss due to evaporation and gassing.
(ii) Use no materials or finishes which will be attacked by acid in the battery room. Spilled acid, and acid vapour given off during gassing, will quickly corrode most exposed metals other than lead. Use an asphalt floor, and coat wooden surfaces liberally with anti-acid paint.
(iii) Ventilate the battery room well, if necessary using corrosion-proofed fans.
(iv) Do not allow a naked flame in the room; and prevent sparking by switching off a circuit before connecting and disconnecting. The gases present are explosive when in the correct proportions.
(v) Mop up spilled acid immediately and wash with a soda solution. Acid on clothing will quickly cause holes to appear.
(vi) Do not allow acid to enter the eyes. If it does so, immediately lie down and run clean water over the eye for as long as possible. Consult a doctor. Acid on the hands is not in itself dangerous, but can be easily transferred to more vulnerable parts of the body.
(vii) When mixing acid for the initial charge of new cells, always add acid to water, and *not* the reverse.
(viii) Watch cell temperature, as excessive heat will damage lead-acid cells. Acid temperature should not exceed 36°C.
(ix) Keep battery terminals clean and coated with petroleum jelly.

Alkaline cells: maintenance

These cells are lighter than the lead-acid type, have greater mechanical strength, can withstand heavy currents without damage, give off no corrosive fumes, and are not affected by being left in the discharged condition. Charge is held much better than with lead-acid cells, the makers claiming 70% capacity after three years without attention. Trickle-charging is thus seldom necessary, although it can be employed when required. Charging after use, or at six-monthly intervals, is all the attention usually required.

The specific gravity of the cells does not change with condition from the value 1·19, the only indication of charge state being terminal voltage, which will normally be about 1·3 V on open circuit or 1·75 V on charge. If in doubt, continue to charge, as these cells are not damaged by overcharging. Constant-voltage or constant-current charging can be used, constant current being about twice the value for a lead-acid cell of the same capacity. Topping up with distilled water is necessary, but precautions (ii) to (vii) for lead-acid cells do not apply. Internal resistance of these cells is generally higher than the equivalent values for lead-acid cells, so voltage varies over a wider range with load.

Comparative charge/discharge characteristics for lead-acid and alkaline cells are shown in Fig. 8.8.

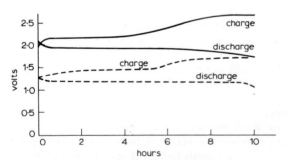

Fig. 8.8 Operational characteristics of cells
——— typical lead-acid cell
----- typical alkaline cell

8.5 INTERNAL RESISTANCE

The path taken by current as it passes through a cell will have resistance. This is the internal resistance of the cell, and is important because of the voltage drop which it causes in the cell, and which results in the terminal voltage being less than the e.m.f. on discharge. The internal voltage drop, given by multiplying internal resistance by current, must be subtracted from the e.m.f. to find the output voltage; that is

$$E - IR_c = V$$

where
E = cell e.m.f., V
I = current taken from the cell, A
R_c = cell internal resistance, Ω
V = cell terminal p.d., V

The internal resistance of a cell usually increases with its age, and with ill treatment such as excessive discharging current, standing in the discharged condition etc.

EXAMPLE 8.1

A cell has an internal resistance of 0·02 Ω and an e.m.f. of 2·2 V. What is its terminal p.d. if it delivers (a) 1 A, (b) 10 A or (c) 50 A?

(a)
$$V = E - IR_c = 2·2 - (1 \times 0·02) \text{ volts}$$
$$= 2·2 - 0·02 \text{ volts}$$
$$= 2·18 \text{ V}$$

(b)
$$V = E - IR_c = 2·2 - (10 \times 0·02) \text{ volts}$$
$$= 2·2 - 0·2 \text{ volts}$$
$$= 2 \text{ V}$$

(c)
$$V = E - IR_c = 2·2 - (50 \times 0·02) \text{ volts}$$
$$= 2·2 - 1·0 \text{ volts}$$
$$= 1·2 \text{ V}$$

This example illustrates how the terminal p.d. falls off as the current increases. The effect is noticeable when a car is started at night. The heavy current taken by the starter motor reduces the battery voltage and the lights dim for as long as the starter is used.

The internal resistance of a cell depends on its design, construction, age and condition. Internal resistance can be measured using a high-resistance voltmeter and an ammeter connected with a switch and resistor (Fig. 8.9). When the switch is open, no current is taken from the cell (if we neglect the voltmeter current), so the voltmeter reads cell e.m.f. If the switch is closed, the terminal voltage and current are measured.

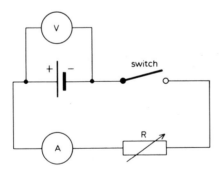

Fig. 8.9 Circuit for calculation of cell internal resistance

Since
$$V = E - IR_c$$
$$R_c = \frac{E - V}{I}$$

The resistor R is used to adjust the cell current to a convenient value which will simplify the calculation.

EXAMPLE 8.2

The e.m.f. of a cell is measured as 2·1 V, and its terminal p.d. as 1·9 V when it carries a current of 5 A. What is its internal resistance?

$$R_c = \frac{E-V}{I}$$

$$= \frac{2\cdot1 - 1\cdot9}{5} \text{ ohm}$$

$$= \frac{0\cdot2}{5} \text{ ohm}$$

$$= 0\cdot04\ \Omega$$

The symbol for a cell is shown in Fig. 8.10, the positive connection being represented by a long line, and the negative by a shorter, thicker line. It is often convenient to show cell internal resistance as a series-connected external resistor.

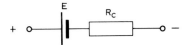

Fig. 8.10 Representation of cell in circuit diagram

When a cell or battery is on charge, the applied terminal voltage must be greater than the e.m.f., so that current is forced against the opposition of the e.m.f. Since the effective voltage is the difference between the terminal voltage and the e.m.f.,

$$I = \frac{V-E}{R_c}$$

from which

$$R_c = \frac{V-E}{I}$$

and

$$V = E + IR_c$$

Compare this with $V = E - IR_c$ for a discharging cell.

EXAMPLE 8.3

A cell with an e.m.f. of 2 V and an internal resistance of $0\cdot08\ \Omega$ is to be charged at 5 A. What terminal voltage must be applied?

$$V = E + IR_c$$
$$= 2 + (5 \times 0\cdot08) \text{ volts}$$
$$= 2 + 0\cdot4 \text{ volts}$$
$$= 2\cdot4\ \text{V}$$

EXAMPLE 8.4

A cell is charged at 10 A when a terminal voltage of 2·7 V is applied. If the cell e.m.f. is 2·2 V, what is the internal resistance?

$$R_c = \frac{V-E}{I}$$

$$= \frac{2\cdot7 - 2\cdot2}{10} \text{ ohm}$$

$$= \frac{0\cdot5}{10} \text{ ohm}$$

$$= 0\cdot05\ \Omega$$

8.6 BATTERIES

A single cell is often incapable of providing a high enough voltage for practical purposes, so a number of cells are connected in series to form a battery. Six cells are shown connected in this way in Fig. 8.11. The total e.m.f. of a battery of cells connected in series is given by multiplying the number of cells by the e.m.f. of each cell, and its internal resistance by multiplying the number of cells by the internal resistance of each cell.

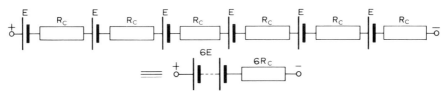

Fig. 8.11 Cells connected in series

EXAMPLE 8.5

(a) A cell of internal resistance 0·05 Ω and e.m.f. 2·2 V is connected to a 0·95 Ω resistor. What current will flow?

(b) What current will flow if the same resistor is connected to a battery of six series-connected cells of this specification?

(a)
$$I = \frac{E}{R + R_c}$$
$$= \frac{2 \cdot 2}{0 \cdot 95 + 0 \cdot 05} \text{ amperes}$$
$$= \frac{2 \cdot 2}{1} \text{ amperes}$$
$$= 2 \cdot 2 \text{ A}$$

(b)
$$\text{Battery e.m.f.} = 6 \times 2 \cdot 2 \text{ volts}$$
$$= 13 \cdot 2 \text{ V}$$
$$\text{Battery internal resistance} = 6 \times 0 \cdot 05 \text{ ohm}$$
$$= 0 \cdot 3 \text{ }\Omega$$

$$I = \frac{E}{R + R_c}$$
$$= \frac{13 \cdot 2}{0 \cdot 95 + 0 \cdot 3} \text{ amperes}$$
$$= \frac{13 \cdot 2}{1 \cdot 25} \text{ amperes}$$
$$= 10 \cdot 6 \text{ A}$$

Notice that the use of a battery of six cells has not increased the current six times.

Sometimes the current which can be provided by each cell for an acceptable fall in terminal voltage will be smaller than that required, and cells are then connected in parallel. It is important to notice that cells must never be parallel-connected unless they are identical in terms of e.m.f. and internal resistance. Discrepancies in e.m.f. will result in internal circulating currents in the battery, or unequal loadsharing, both these faults causing rapid deterioration of healthy cells. The e.m.f. of a parallel-connected battery is that of each cell, whereas its internal resistance is the resultant resistance of the individual internal resistances connected in parallel. The arrangement is shown in Fig. 8.12.

EXAMPLE 8.6

A cell of e.m.f. 1·6 V and internal resistance of 0·3 Ω is connected to a 0·1 Ω resistor. What current flows? Find the current if six of these cells are connected in parallel to the same load.

$$I = \frac{E}{R + R_c}$$

$$= \frac{1\cdot6}{0\cdot1 + 0\cdot3} \text{ amperes}$$

$$= \frac{1\cdot6}{0\cdot4} \text{ amperes}$$

$$= 4\text{ A}$$

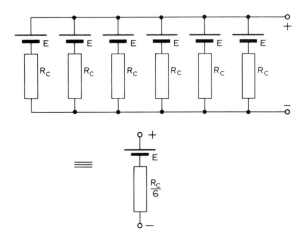

Fig. 8.12 Cells connected in parallel

For the battery, $\qquad E = 1\cdot6\text{ V}$

and internal resistance $\qquad R_c = \dfrac{0\cdot3}{6}$ ohms

$$= 0\cdot05\,\Omega$$

Therefore $\qquad I = \dfrac{E}{R + R_c}$

$$= \frac{1\cdot6}{0\cdot1 + 0\cdot05} \text{ amperes}$$

$$= \frac{1\cdot6}{0\cdot15} \text{ amperes}$$

$$= 10\cdot7\text{ A}$$

Where high voltage *and* high current are necessary, series-parallel arrangements of cells are used, connected as shown in Fig. 8.13. All cells must be identical, and the e.m.f. will be found by multiplying the e.m.f. per

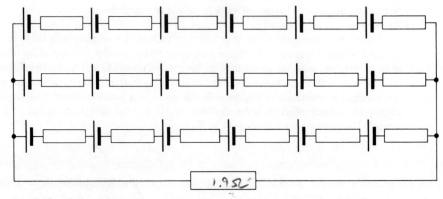

Fig. 8.13 Cells connected to load in series-parallel

cell by the number of cells connected in series. Overall internal resistance will be given by the expression

$$\frac{\text{resistance of each cell} \times \text{number of cells in series}}{\text{number of parallel groups}}$$

EXAMPLE 8.7

Eighteen cells, each of e.m.f. 2·4 V and internal resistance 0·05 Ω, are connected in three banks of six cells in series. The three banks are then connected in parallel with each other and with a resistor of 1·9 Ω. Find the current flow in the resistor.

The arrangement is shown in Fig. 8.13.

$$\text{Battery e.m.f.} = 6 \times 2 \cdot 4 \text{ volts}$$
$$= 14 \cdot 4 \text{ V}$$
$$\text{Battery internal resistance} = \frac{0 \cdot 05 \times 6}{3} \text{ ohm}$$
$$= 0 \cdot 1 \, \Omega$$
$$I = \frac{E}{R + R_c}$$
$$= \frac{14 \cdot 4}{1 \cdot 9 + 0 \cdot 1} \text{ amperes}$$
$$= \frac{14 \cdot 4}{2} \text{ amperes}$$
$$= 7 \cdot 2 \text{ A}$$

8.7 CAPACITY AND EFFICIENCY

A battery of cells is a device for storing energy, the energy stored being known as the **capacity** of the battery. This is usually measured in **ampere hours** (Ah). Energy is given not, in fact, by multiplying together current and time, but by current × time × voltage, so this method of measurement is not strictly correct. If, however, a battery is capable of providing a current of 5 A for 10 h, it is said to have a capacity of 50 Ah. Such a battery could not be expected to provide, say, a current of 10 A for 5 h, since capacity decreases as the current taken increases. Capacity is therefore based on a definite discharge time, usually 10 h when the capacity is quoted **at the 10 h rate**.

Losses occur both in converting electrical energy to chemical energy (charging) and in the reverse operation of discharging. No cell can thus be 100% efficient, the actual efficiency being given, as usual, by

$$\text{efficiency} = \frac{\text{output}}{\text{input}} \times 100\%$$

There are two methods of measuring the efficiency of a cell.

$$\text{Ampere-hour efficiency} = \frac{\text{output (discharge), Ah} \times 100\%}{\text{input (charge), Ah}}$$

A cell in good condition is likely to have an ampere-hour efficiency in the region of 80%, which is high since it assumes that charging and discharging terminal voltages are the same. In fact, charging terminal voltage is the sum of cell e.m.f. and internal voltage drop, whereas discharging terminal-voltage drop is given by their difference. A theoretically truer method of calculating efficiency is the watt hour method.

$$\text{Watt-hour efficiency} = \frac{\text{average discharge, watt hours} \times 100\%}{\text{average charge, watt hours}}$$

Watt-hour efficiency for a cell in good condition is likely to be in the region of 65%.

EXAMPLE 8.8

A discharged 12 V battery is charged for 10 h at 12 A, the average charging terminal voltage being 15 V. When connected to a load, a current of 10 A for 9 h at an average terminal voltage of 12 V discharges the battery. Calculate (*a*) the ampere-hour efficiency, and

(b) the watt-hour efficiency.

(a)
$$\text{Ampere-hour efficiency} = \frac{10 \times 9 \times 100\%}{10 \times 12}$$
$$= 75\%$$

(b)
$$\text{Watt-hour efficiency} = \frac{10 \times 9 \times 12 \times 100\%}{10 \times 12 \times 15}$$
$$= 60\%$$

8.8 SUMMARY OF FORMULAS FOR CHAPTER 8

Discharging:
$$V = E - IR_c \qquad E = V + IR_c \qquad R_c = \frac{E - V}{I}$$

Charging:
$$V = E + IR_c \qquad E = V - IR_c \qquad R_c = \frac{V - E}{I}$$

where
V = cell or battery terminal voltage, V
E = cell or battery e.m.f., V
R_c = cell or battery internal resistance, Ω
I = charge or discharge current, A

For n identical cells in series,

internal resistance = nR_c, where R_c = internal resistance of each cell

e.m.f. = nE, where E = e.m.f. of each cell

For m identical cells in parallel,

$$\text{internal resistance} = \frac{R_c}{m}$$

$$\text{e.m.f.} = E$$

For a group of m parallel sets, each of n cells in series,

$$\text{internal resistance} = \frac{nR_c}{m}$$

$$\text{e.m.f.} = nE$$

8.9 EXERCISES

1. Sketch one each of any form of (a) primary cell, and (b) secondary cell. Label the component parts clearly. Describe how each cell operates, and the e.m.f. in each case. (C & G)

2. (a) Describe a nickel-cadmium (alkaline) cell. Give the characteristic charge and discharge curves, and discuss briefly the advantages and disadvantages of this form of secondary cell.
   (b) Make a sketch of any one form of primary cell, labelling the separate parts. (C & G)

3. Make a labelled sketch showing a section through a single-cell dry battery commonly used in hand torches. (Part of C & G)

4. A cell of e.m.f. 1·5 V has an internal resistance of 0·2 Ω. Calculate its terminal p.d. if it delivers a current of 0·5 A.

5. A cell of e.m.f. 2·2 V and internal resistance 0·05 Ω has a 1·05 Ω resistor connected across its terminals. Calculate the current flow and the terminal p.d. of the cell.

6. The e.m.f. of a cell is measured with a high-resistance voltmeter on open circuit, and is found to be 1·45 V. When a current of 1 A is drawn from the cell, the terminal p.d. falls to 1·25 V. What is the internal resistance of the cell?

7. Describe a lead-acid secondary cell. Explain briefly the changes in the cell during charge and discharge.
   The potential difference between the terminals of a lead-acid cell on open circuit was 2·18 V; when the cell was discharging at the rate of 9 A the terminal p.d. was 2·02 V. Calculate the internal resistance of the cell. (C & G)

8. Describe, with sketches, a lead-acid secondary cell and state briefly the chemical changes in the cell during charge and discharge. (The chemical formulas are *not* required). Explain the importance of a low internal resistance. A lead-acid cell discharging at the rate of 6 A has a terminal p.d. of 1·95 V. On open circuit, the p.d. is 2·1 V. Calculate the internal resistance of the cell. (C & G)

9   The electrolyte in a lead-acid cell is .... (ULCI)

10 A battery consists of six 2 V cells in series. Calculate the e.m.f. of the battery.

11 Six cells, each of e.m.f. 1·5 V and internal resistance 0·2 Ω, are connected in series to a 1·8 Ω resistor. Calculate the current delivered, and the battery terminal voltage on load.

12 (a) What is the total e.m.f. when a number of cells are connected in series?
    (b) What is the purpose of connecting a number of cells in parallel? (NCTEC)

13 A lead-acid battery for an electric truck has 15 series-connected cells, each with an e.m.f. of 2·3 V and an internal resistance of 0·01 Ω. Calculate the terminal voltage when the battery delivers a current of 50 A.

14 A cell has an open-circuit terminal voltage of 2·1 V and an internal resistance of 0·1 Ω. Calculate the terminal p.d. of the cell on charge when the charging current is (a) 2 A, (b) 10 A.

15 (a) State the difference between a primary and a secondary cell.
    (b) A battery comprises five primary cells, each cell having an e.m.f. of 1·1 V and a rated current of 2 A. Calculate the e.m.f. and the rated current of the battery when the cells are connected (i) in series, (ii) in parallel. Draw circuit diagrams for (i) and (ii). (ULCI)

16 A battery is made up of 12 identical cells connected in series. Each cell has an e.m.f. of 2 V and an internal resistance of 0·05 Ω. What terminal voltage must be applied to the battery if a charging current of 20 A is required?

17 A battery with an e.m.f. of 12 V charges at 10 A when 16 V is applied to it. What is the internal resistance of the battery?

18 A cell with an internal resistance of 0·15 Ω has a terminal p.d. of 1·8 V when charging at 5 A. What is the e.m.f. of the cells?

19 A lead-acid battery comprises 50 cells in series, each of open-circuit e.m.f. 2 V and internal resistance 0·02 Ω. Calculate the terminal voltage
    (a) when supplying a load of 10 A
    (b) when being charged at 10 A. (C & G)

20 Three cells, each of e.m.f. 1·4 V and internal resistance 0·3 Ω, are connected in parallel to a 0·9 Ω resistor. Calculate the current in the resistor and the battery terminal p.d.

21 Twelve lead-acid cells, each of e.m.f. 2·1 V and internal resistance 0·015 Ω, are connected in three series banks of four cells. The banks are connected in parallel to a load resistor. If a current of 20 A flows in this resistor, calculate
    (a) the resistor value,
    (b) the terminal p.d. of the battery.

22 A battery of nine primary cells is connected
    (a) all cells in series;
    (b) all cells in parallel;
    (c) three sets in parallel, each set consisting of 3 cells in series.
    Each cell has an e.m.f. of 1·4 V and an internal resistance of 0·45 Ω. The battery terminals are connected to a circuit of resistance 7·2 Ω. Calculate in each case (i) the current in the 7·2 ohm resistance, (ii) the voltage drop across the resistance. (C & G)

23 A discharged lead-acid battery is charged at 5 A for 15 h at an average voltage of 7·2 V. On discharge, the battery gives 6 A for 10 h at an average p.d. of 6 V. Calculate the ampere-hour and watt-hour efficiencies.

Chapter 9

Electromagnetic induction

9.1 INTRODUCTION

For many hundreds of years, scientists knew of the existence of the electric current and of the magnetic field, but the two were considered to have no connection. Thanks to the work of Oersted, Faraday and others, the two are now considered to be inseparable. In Chapter 6, it was shown that the flow of current in a conductor gave rise to a magnetic field. In this chapter, it will be shown that, under certain conditions, a magnetic field can be responsible for the flow of an electric current.

This effect is known as electromagnetic induction.

9.2 DYNAMIC INDUCTION

The word 'dynamic' suggests force and movement; in dynamic induction, a conductor is moved through a **magnetic field**.

If a length of flexible conductor has its ends connected to a sensitive indicating instrument, the needle of the instrument will give a sharp 'kick' when the conductor is suddenly moved in a magnetic field (Fig. 9.1).

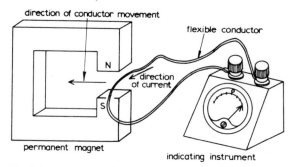

Fig. 9.1 Induced e.m.f. in conductor due to its movement in magnetic field

This simple experiment shows that, if a conductor moves in a magnetic field, an e.m.f. is induced in it, and this e.m.f. will cause a current to flow if a closed electric circuit exists. If a number of conductors are assembled so that they can rotate between the poles of a magnet, we have a simple generator, the principles of which will be further explained at a later stage.

If the flexible conductor is wound in a loop, so that the adjacent sides of two turns pass at the same speed through the same magnetic field, the deflection of the instrument will be twice as great as when the single conductor was used. From this, it follows that induced e.m.f. depends on the length of conductor subject to the magnetic field.

If the magnet is changed for a stronger one (with greater magnetic-flux density between its polefaces), the deflection will again be greater. Thus induced e.m.f. depends on the density of the magnetic field through which the conductor passes.

The deflection of the needle can be shown to depend on the speed, or velocity, of the conductor through the magnetic field. The faster the conductor moves, the greater is the deflection.

These three effects can be shown by experiment to be related by the formula

$$e = Blv$$

where
e = induced e.m.f., V
B = flux density of magnetic field, T
l = length of conductor in the field, m
v = velocity (speed) of the conductor, m/s

It should be noted that e will be a steady value only as long as the conductor velocity is constant, and as long as the flux density of the magnetic field remains the same. For this reason, the symbol e, for an instantaneous value, is used. The conductor must move directly across the magnetic field, so that its path is at right angles to the magnetic flux.

EXAMPLE 9.1

At what velocity must a conductor, 0·1 m long, cut a magnetic field of flux density 0·8 T if an e.m.f. of 4 V is to be induced in it?

$$e = Blv, \quad \text{so} \quad v = \frac{e}{Bl}$$

$$v = \frac{4}{0.8 \times 0.1} \text{ metres per second}$$

$$= 50 \text{ m/s}$$

It should be noted that, whenever a conductor moves in a magnetic field, it will have an e.m.f. induced in it, but that this e.m.f. can only result in a current if there is a closed circuit. The value of such a current will depend on the e.m.f. and the circuit resistance, following the relationship $I = E/R$, provided that E is constant.

EXAMPLE 9.2

A conductor 300 mm long moves at a uniform speed of 2 m/s through a uniform magnetic field of flux density 1 T. What current will flow in the conductor
    (a)   if its ends are open-circuited?
    (b)   if its ends are connected to a 12 Ω resistor?

In both cases:

$$e = Blv$$
$$= 1 \times 0.3 \times 2 \text{ volts}$$
$$= 0.6 \text{ V}$$

(a) If the ends of the conductor are open-circuited, no current will flow.

(b)
$$I = \frac{E}{R}$$
$$= \frac{0.6}{12} \text{ ampere}$$
$$= 0.05 \text{ A} \quad \text{or} \quad 50 \text{ mA (neglecting conductor resistance)}$$

EXAMPLE 9.3

A conductor of effective length 0·5 m is connected to a milliammeter of resistance 1 Ω, which reads 15 mA when the conductor moves at a steady speed of 40 mm/s in a magnetic field. What is the average flux density of the field?

$$E = IR$$
$$= 0.015 \times 1 \text{ volt}$$
$$= 0.015 \text{ V}$$

$$e = Blv, \quad \text{so} \quad B = \frac{e}{lv}$$

Therefore
$$B = \frac{0.015}{0.5 \times 0.04} \text{ tesla}$$
$$= 0.75 \text{ T}$$

An alternative method of calculating induced e.m.f. is based on the definition of the unit of magnetic flux, the weber. If a conductor moving at constant speed cuts a total of one weber of flux in one second, the e.m.f. induced in it will be one volt. Thus the induced e.m.f. is equal to the average rate of cutting magnetic flux in webers per second; that is

$$\text{volts} = \text{webers/second}$$

$$\text{or} \quad e = \frac{\Phi}{t}$$

EXAMPLE 9.4

A conductor is moved through a magnetic field, having a total flux of 0·2 Wb, in 0·5 s. What will be the average e.m.f. induced?

$$e = \frac{\Phi}{t}$$

$$= \frac{0\cdot 2}{0\cdot 5} \text{ volt}$$

$$= 0\cdot 4 \text{ V}$$

9.3 RELATIVE DIRECTIONS OF E.M.F., MOVEMENT AND FLUX

If the direction of movement of the conductor in the magnetic field (Figs. 9.1 and 9.2) is reversed, the e.m.f. induced will have reversed polarity. If the magnetic-field direction is reversed, this will again reverse the polarity.

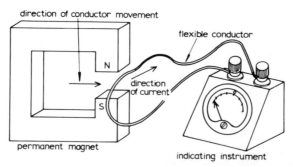

Fig. 9.2 Dependence of induced e.m.f. direction on direction of conductor movement relative to magnetic field

It is often necessary to be able to forecast the direction of one of the three variables (magnetic field, induced e.m.f., and conductor movement) if the other two are known. The easiest of several methods is the application of **Fleming's right-hand (generator) rule**.

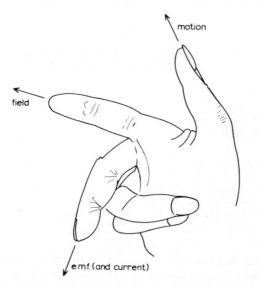

Fig. 9.3 Position of right hand for application of Fleming's right-hand rule

The thumb and first two fingers of the *right* hand are held mutually at right angles (Fig. 9.3). Then, if the **F**irst-finger direction is that of the magnetic **F**ield (north pole to south pole outside the magnet), the se**C**ond-figure direction is that of **C**urrent as a result of induced e.m.f.; then the thu**M**b shows the direction of conductor **M**ovement in the magnetic field. It is most important to ensure that the *right* hand is used.

EXAMPLE 9.5

Examine the diagrams of a conductor between a pair of magnetic poles (Fig. 9.4), and state

(i)   the direction in which the conductor moves for Fig. 9.4a and b
(ii) the polarity of the magnet system for Fig. 9.4c
(iii) the direction of the induced e.m.f. for Fig. 9.4d.

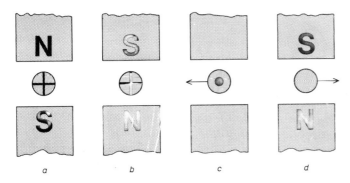

Fig. 9.4 Diagram for example 9.5

The answers are

   (a) left to right
   (b) right to left
   (c) north pole at the top
   (d) upwards (out of the paper).

9.4 SIMPLE ROTATING GENERATOR

We have already seen that, when a conductor moves across a magnetic field, it has an e.m.f. induced in it. When the conductor moves *along* the field, no flux is cut, and no e.m.f. induced. If the conductor moves at right angles to the field, flux is cut at the maximum rate and maximum e.m.f. induced. If the conductor cuts the field obliquely, the e.m.f. induced lies between the zero and maximum values, and depends on the angle between the line taken by the conductor and the magnetic flux.

Consider a simple rectangular loop which can be rotated between the poles of a magnet system as shown in Fig. 9.5. Connections are made to the ends of the loop by means of brushes bearing on slip rings. A section

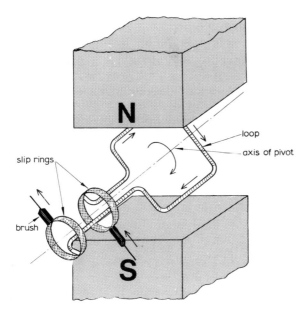

Fig. 9.5 Loop connected to slip rings and able to rotate in magnetic field

of the arrangement is shown in Fig. 9.6a. When a conductor is in either of positions 1 and 5, it is moving along the lines of magnetic flux, but not cutting them, and so no e.m.f. is induced in it. In positions 2, 4, 6 and 8, the conductor cuts the magnetic flux at an angle, so an e.m.f. is induced. At positions 3 and 7, the conductor moves directly across the magnetic flux, cutting it at a maximum rate, and a maximum e.m.f. is induced. Application of Fleming's right-hand rule will show that, as the conductor moves from left to right

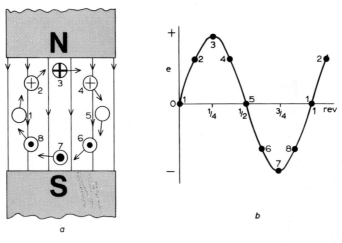

Fig. 9.6 *a* Section through loop of Fig. 9.5, which is shown in several positions
b E.M.F. in loop drawn as graph

through the magnetic field (positions 2, 3 and 4), the direction of the induced e.m.f. is the opposite of that induced when the conductor passes from right to left through the magnetic field (positions 6, 7 and 8). The e.m.f. induced in the conductor is drawn as a graph to a base of time in Fig. 9.6b. From this, we can see that the e.m.f. is **alternating**, that is alternately acting in opposite directions.

The opposite directions of the e.m.f.s in opposite loop sides will drive a current through the external circuit. This current will alternate, reversing as the loop sides pass through positions 1 and 5 of Fig. 9.6, and reaching maximum as the loop sides pass through positions 3 and 7.

This generator produces alternating current, and is known as an **alternator**. Alternators are used in power stations to provide electricity for the grid system. These very large machines have a different form of construction to our simple loop generator, but the principle is the same.

9.5 DIRECT-CURRENT GENERATOR

If we require a direct-current output, as we do with a d.c. generator, we must wire a changeover switch into circuit, and operate it to change the polarity each time the e.m.f. reaches zero. The resulting e.m.f. would vary continuously, but its polarity would not change. It would be physically impossible to operate such a switch by hand, because the generator revolves too quickly. What is required is an automatic switch.

Such a switch is in the form of a conducting cylinder, cut along its length and with the two halves insulated from each other. The simple **commutator** thus formed is mounted on the shaft of the machine (Fig. 9.7a). A simple loop of wire forms a suitable conductor system, the two ends of the loop being connected to the two halves of the commutator. Stationary brushes rest on the commutator surface, and make contact with it, the output from this simple-loop generator being shown in Fig. 9.7b.

Fig. 9.8 shows the operation of the generator in simple diagrams. Fig. 9.8a shows the two conductors approaching the point of maximum induced e.m.f. Since the individual conductor e.m.f.s are in different directions, they add up round the loop to give a total e.m.f. equal to twice that of each conductor. Current flows to the external load, which is connected to the loop via the commutator and brushes. In Fig. 9.8b, the two conductors are approaching the point of zero e.m.f., and the brushes are approaching their point of changeover from one commutator segment to the next. In Fig. 9.8c, the changeover has occurred and the lower circuit conductor is connected to rotating conductor 1, and not to conductor 2 as previously. The conductor 1 is, however, travelling through the magnetic field in a direction opposite to that which it had in Fig. 9.8a, so its e.m.f. has reversed to be the same as that previously induced in conductor 2. The direction of current in the external circuit is thus unchanged. The commutator then ensures that whichever conductor is passing the north pole is always connected to the negative end of the load, and the conductor passing the south pole to the positive end.

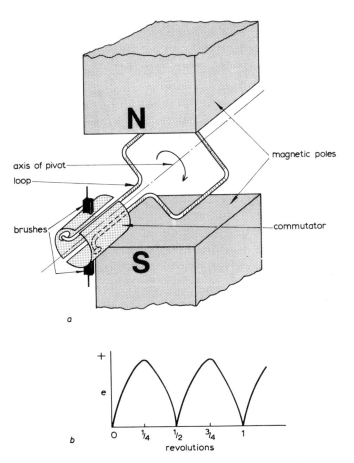

Fig. 9.7 *a* Loop connected to simple commutator and able to rotate in magnetic field
 b Output e.m.f. from system

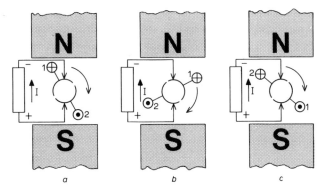

Fig. 9.8 Principle of d.c. generator

Practical d.c. generators have many conductor loops, and commutators with large number of segments, so the output is almost steady. The magnetic flux cut by the conductors must be as great as possible to ensure maximum induced e.m.f., and this is arranged by means of a **magnetic circuit**. This circuit consists of the iron polepieces, the iron body of the rotor and the frame of the machine, which is called the **yoke**. A d.c. machine may have any even number of poles (2, 4, 6, 8 etc.). The loops in which the e.m.f. is induced; referred to as the **armature winding**, are let into slots in the surface of the iron cylinder forming the rotor, or rotating part of the machine. The magnetic flux is set up by field windings, which are solenoids, placed on the salient poles.

9.6 STATIC INDUCTION

The induced e.m.f. (Section 9.2) was due to movement. We considered the conductor to move in a stationary field, but, of course, a similar e.m.f. would be induced if the conductor remained still and the field moved past it. How an e.m.f. can be induced without any physical motion at all is described here.

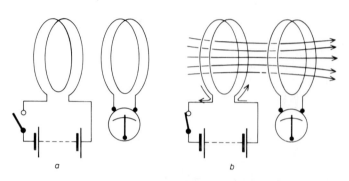

Fig. 9.9 Static induction

Consider two coils of wire placed side by side (Fig. 9.9), but not touching or in electrical contact with each other. The first coil is connected in series with a battery and a switch, so that a current can be made to flow in it and can then be switched off. The second coil has a measuring instrument connected to its ends.

If the switch in the circuit of the first coil is operated, the instrument connected to the second coil is seen to 'kick' and then return to zero. This happens each time the switch is turned on or off, the needle moving in a different direction at each operation. A reference to Fig. 9.9 shows the reason for the induced e.m.f. In Fig. 9.9a, the switch is off, and the first coil sets up no magnetic flux. When the switch is on (Fig. 9.9b), the first coil sets up a magnetic flux, some of which passes through, or 'links' with, the second coil. There has been a change in the flux linking the second coil, which has had an e.m.f. induced in it, just as if the coil had moved into a steady magnetic field. The e.m.f. will only be induced while the magnetic flux is changing. When the flux becomes steady, no e.m.f. is induced.

If the left-hand coil is fed from a source of alternating current, the magnetic flux set up will be continually changing and an alternating e.m.f. will be induced in the right-hand coil. This is the principle of the transformer, which will be considered in Chapter 10.

A further study of Fig. 9.9 will show that a change of magnetic flux linkages has taken place in the left-hand coil, as well as in the right-hand coil. The left-hand coil, like the right-hand coil, will thus have an e.m.f. induced in it, but this e.m.f. will oppose the battery voltage and try to slow the change of current. This **self-induced e.m.f.** is sometimes called a **back e.m.f.**, and any circuit which has the property of inducing such an e.m.f. in itself is said to be **self-inductive**, or just **inductive**. All circuits are to some extent, self-inductive, but some conductor arrangements give rise to a much greater self inductance than others. The unit of self inductance, which has the symbol L, is the **henry** (symbol H). The property of self inductance is discussed fully in *Electrical craft principles* – Vol. 2.

9.7 SUMMARY OF FORMULAS FOR CHAPTER 9

$$e = Blv \qquad B = \frac{e}{lv} \qquad l = \frac{e}{Bv} \qquad v = \frac{e}{Bl}$$

where
- e = induced e.m.f., V
- B = flux density of magnetic field, T
- l = length of conductor in the field, m
- v = velocity (speed) of the conductor, m/s

$$e = \frac{\Phi}{t} \qquad \Phi = et \qquad t = \frac{\Phi}{e}$$

where
- e = average induced e.m.f., V
- Φ = total magnetic flux cut, Wb
- t = time taken to cut flux, s

Fleming's right-hand (generator) rule

- First finger points direction of magnetic flux
- second finger points direction of induced e.m.f.
- thumb points direction of conductor movement through the magnetic field.

9.8 EXERCISES

1. Examine the diagrams in Fig. 9.10 which show a conductor being moved in a magnetic field, and state
   (a) the direction of induced e.m.f. for Fig. 9.10a
   (b) the direction of induced e.m.f. for Fig. 9.10b
   (c) the magnetic polarity for Fig. 9.10c
   (d) the direction of conductor movement for Fig. 9.10d.

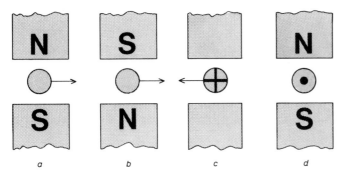

Fig. 9.10 Diagrams for exercise 9.8 (1)

2. A conductor is moved at a speed of 10 m/s directly across a magnetic field of flux density 15 mT, and has an e.m.f. of 0·3 V induced in it. What is its effective length?

3. At what speed must a conductor of effective length 180 mm be moved at right angles to a magnetic field of flux density 0·6 T to induce in it an e.m.f. of 0·324 V?

4. A conductor of effective length 200 mm connected to a milliammeter of resistance 5 Ω is moved through a magnetic field of flux density 0·5 T. If the milliammeter reads 40 mA, at what speed must the conductor be moving?

5. What e.m.f. will be induced in a conductor of effective length 80 mm which is moving with a velocity of 15 m/s through a magnetic field of flux density 0·4 T?

6. One conductor of a generator is 500 mm long and moves at a uniform velocity of 20 m/s in the pole flux which has an average density of 0·4 T. What is the average e.m.f. induced in the conductor? If the winding has 200 of these conductors connected in series, what is the total generated e.m.f.?

7. A conductor is subjected to a magnetic flux changing at the rate of 4 Wb/s. What e.m.f. is induced in the conductor?

8. An average e.m.f. of 1·5 V is induced in a conductor while the initial linking flux of 0·25 Wb is falling to zero. How long does the flux take to collapse?

9. A millivoltmeter connected to a conductor reads a steady 20 mV for 3 s while the conductor is subjected to a changing magnetic flux. Calculate the total flux change.

10. When a magnet is being inserted into a coil of wire, what factors govern
    (a) the direction,
    (b) the magnitude of the induced e.m.f.? (NCTEC)

11. Describe with the aid of a sketch a simple-loop generator which consists of a single wire loop rotating between the poles of a permanent magnet. Show the output of the loop taken from slip rings, and sketch a graph of the output voltage to a base of time.

12. Draw a diagram to show a simple 2-part commutator which can be substitiued for the slip rings of the generator of exercise 11. Describe how the commutator functions, and sketch a graph of the output from the machine.

13. Describe, with the aid of a sketch, the construction and action of a simple direct-current generator.
    State
    (a) the factors on which the generated e.m.f. depends
    (b) how the generated e.m.f. can be controlled. (ULCI)

Chapter 10
Basic alternating current theory

10.1 WHAT IS ALTERNATING CURRENT?

Although direct-current systems and calculations are still indispensable to the electrical engineer, most public supplies are now alternating-current mains. The reasons for the changeover from d.c. to a.c. supplies will be considered in Section 10.2, our purpose here being to indicate how the two systems differ. The easiest method of portraying an alternating quantity is to draw a graph showing how it varies with time, as in Fig. 10.1. Any part of the graph which lies above the horizontal (or zero) axis represents current or

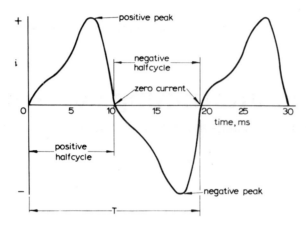

Fig. 10.1 Graph of alternating current

voltage in one direction, and values below it represent current or voltage in the other direction. The pattern given by the graph is known as the **waveform** of the a.c. system, and this usually repeats itself. There is no need for the waveform above the zero axis to have the same shape as that below it, although in most a.c. systems derived from mains supplies, this is the case.

An alternating current is thus one which rises in one direction to a maximum value, before falling to zero and repeating in the opposite direction. Instead of drifting steadily in one direction, the electrons forming the current move backwards and forwards in the conductor.

The time taken for an alternating quantity to complete its pattern (to flow in both directions and then to return to zero) is called the **periodic time** (symbol T) for the system, which is said to complete **one cycle** in this time. The complete cycle is split into the **positive halfcycle** above the axis, and the **negative halfcycle** below it.

The number of complete cycles traced out in a given time is called the frequency (symbol f), usually expressed in **hertz** (Hz), which are **cycles per second** (c/s). If there are f cycles in one second, each cycle takes $1/f$ second, so that

$$T = \frac{1}{f} \quad \text{and} \quad f = \frac{1}{T}$$

EXAMPLE 10.1

Calculate the frequency of the a.c. system shown in Fig. 10.1.

From Fig. 10.1,
$$T = 20\,\text{ms} = 0\cdot02\,\text{s}$$

Therefore
$$f = \frac{1}{T}$$
$$= \frac{1}{0\cdot02}\,\text{hertz}$$
$$= 50\,\text{Hz}$$

A frequency of 50 Hz is the standard for the supply system in many parts of the world, including the UK, but 60 Hz systems are also common for mains supplies.

10.2 ADVANTAGES OF A.C. SYSTEMS

There are certain complications which occur when using a.c. supplies, which are absent with d.c. supplies; these complications are explained later in this chapter. However, the advantages of a.c. supplies have led to their general use, some of the more important being

(a) An alternating-current generator (often called an alternator) is more robust, less expensive, requires less maintenance, and can deliver higher voltages than its d.c. counterpart.

(b) The power loss in a transmission line depends on the square of the current carried ($P = I^2R$). If the voltage used is increased, the current is decreased, and losses can be made very small. The simplest way of stepping up the voltage at the sending end of a line, and stepping it down again at the receiving end, is to use transformers, which will only operate efficiently from a.c. supplies.

(c) 3-phase a.c. induction motors are cheap, robust and easily maintained.

(d) Energy meters, to record the amount of electricity used, are much simpler for a.c. supplies than for d.c. supplies.

(e) Discharge lamps (fluorescent, sodium, mercury vapour etc.) operate more efficiently from a.c. supplies, although filament lamps are equally effective on either type of supply.

(f) Direct-current systems are subject to severe corrosion, which is hardly present with a.c. supplies.

10.3 VALUES FOR A.C. SUPPLIES

The alternating current or voltage changes continuously, so that it is not possible to state its value in the same simple terms that can be used for a direct current.

Instantaneous values are the values at particular instants of time, and will be different for different instants. Symbols for instantaneous values are small symbols, v for voltage, i for current, and so on.

Maximum or peak values are the greatest values reached during alternation, usually occurring once in each halfcycle. Maximum values are indicated by V_m for voltage, I_m for current, and so on.

Average or mean value is the average value of the current or voltage. If an average value is found over a full cycle, the positive and negative halfcycles will cancel out to give a zero result if they are identical. In such cases, it is customary to take the average value over a halfcycle. The average value of this kind of waveform can be found as shown in example 10.2. Symbols used are V_{av} for voltage, I_{av} for current, and so on.

EXAMPLE 10.2

The following table gives the waveform of a halfcycle of alternating voltage.

Time, ms	0	0·25	0·5	0·75	1·0	1·25	1·5	1·75
Volts, V	0	45	72	91	104	118	142	185
Time, ms	2·0	2·25	2·5	2·75	3·0	3·25	3·5	
Volts, V	240	278	295	300	280	248	195	
Time, ms	3·75	4·0						
Volts, V	85	0						

Find the frequency of the supply, its instantaneous values after 1·8 ms and 2·4 ms, the maximum value and the mean value of voltage.

$$f = \frac{1}{T}$$

$$= \frac{1}{8\,\text{ms}}$$

$$= \frac{1}{0.008}\,\text{hertz}$$

$$= 125\,\text{Hz}$$

The next step in the solution is to draw the halfcycle as a graph (Fig. 10.2), reading off the instantaneous values (195 V at 1·8 ms, 287 V at 2·4 ms) and its maximum value (300 V).

To find the average or mean value, the base line (time axis) is divided into any number of equal parts. For clarity, eight parts have been chosen, although more would give greater accuracy. At the *centre* of each part, a dotted line has been drawn up to the curve.

The average value of voltage will be the average length of these lines (expressed in volts). To find this, we add the voltage represented by each line and divided by the number of lines.

$$V_{av} = \frac{45 + 91 + 118 + 185 + 278 + 300 + 248 + 85}{8} \text{ volts}$$

$$= \frac{1350}{8} \text{ volts}$$

$$= 169 \text{ V}$$

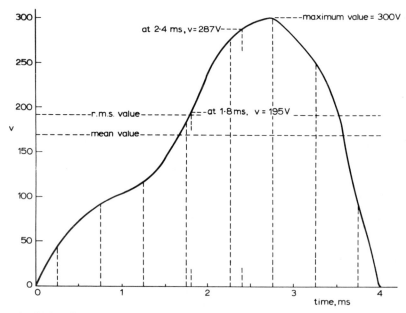

Fig. 10.2 Graph for examples 10.2, 10.3 and 10.4

Effective value

Since the heat dissipated by a current is proportional to its square ($P = I^2R$), the average value of an alternating current is not the same as the direct current which produces the same heat or does the same work in the same time. The equivalent to a direct current is the value we use most in describing and calculating a.c. systems, and is called the **effective** or **root-mean square** (r.m.s.) value of the system. The r.m.s. value is the square root of the average value of the squares of the instantaneous values. The symbols used for r.m.s. values are V, I etc. The method of finding the effective or r.m.s. value of a given waveform is illustrated in the following example.

EXAMPLE 10.3

Find the r.m.s. value of the voltage waveform of example 10.2.

To find the root-mean-square value,

    (a)   divide the base into equal parts and erect a vertical line to the curve from the centre of each part (as for finding average value)
    (b)   square the value of each vertical line
    (c)   take the mean of the squared values (add them and divide by the number of lines)
    (d)   take the square root of the result — this is the root of the mean of the squared values.

The graph has already been drawn and vertical lines have been erected for example 10.2, and need not be repeated in this case. The sum of the squared values will be

$$45^2 + 91^2 + 118^2 + 185^2 + 278^2 + 300^2 + 248^2 + 85^2 = 294\,468 \text{ V}^2$$

$$\text{The mean of the square values} = \frac{294\,468}{8} = 36\,809 \text{ V}^2$$

$$\text{Root-mean-square value} = \sqrt{36\,809} \text{ V} = 191 \cdot 9 \text{ V}$$

It will be seen that the r.m.s. value is greater than the mean value, and this is always the case, except for a direct current, for which they are equal.

Form factor for a particular waveform is the ratio of the r.m.s. and mean values.

$$\text{Form factor} = \frac{\text{r.m.s. value}}{\text{mean value}}$$

EXAMPLE 10.4

Find the form factor of the waveform of example 10.2.

$$\text{Form factor} = \frac{\text{r.m.s. value}}{\text{mean value}}$$

$$= \frac{191 \cdot 9 \text{ V}}{169 \text{ V}}$$

$$= 1 \cdot 136$$

Form factor is an indication of the shape of a waveform; the higher its value the more 'peaky' is the waveshape.

10.4 SINUSOIDAL WAVEFORMS

In Chapter 9, we considered a simple rectangular loop of wire rotating on an axis between the poles of a permanent magnet (Figs. 9.5 and 9.6). The e.m.f. induced in the loop is shown again in Fig. 10.3, one cycle being induced for each revolution of the loop. If the loop rotates at a constant speed, the horizontal axis can be divided into equally spaced units of time as well as degrees of rotation.

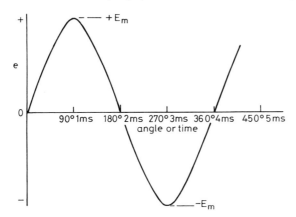

Fig. 10.3 Sinusoidal waveform

The e.m.f. induced in the loop at any instant depends on the rate of cutting lines of magnetic flux. Referring to Fig. 10.4, movement from O to A induces no e.m.f., whereas movement from O to B induces a maximum e.m.f., which we will call E_m. Moving the same distance from O to C at an angle of ϕ degrees to OA will induce an e.m.f. proportional to the number of lines of magnetic-flux cut, that is proportional to the length

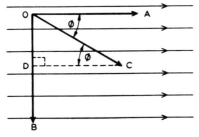

Fig. 10.4 Effect of conductor direction on induced e.m.f.

OD. OD and OC are the opposite side to ϕ and the hypotenuse, respectively, of the right-angled triangle OCD. From simple trigonometry, e.m.f. induced in moving from O to C

$$= E_m \times \frac{OD}{OC}$$

$$= E_m \sin \phi$$

The waveshape of Fig. 10.3 is thus a graph of $E_m \sin \phi$, is referred to as a 'sine wave', and is said to be 'sinusoidal' in shape. This waveshape is easily expressed as a mathematical formula, and is similar to that obtained from practical generators, so from now on we will consider all a.c. electrical systems to have sinusoidal waveshapes.

It is common to use the **radian** as a measure of angle; it is defined as the angle subtended at the centre of a circle by a section of the circumference of equal length to its radius. Since the circumference of a circle is $2\pi \times$ radius, there are 2π radians in $360°$, so 1 radian $= 360°/(2\pi) = 57\cdot3°$ approximately.

The total angular movement after t seconds of a wire loop rotating at f revolutions per second and giving an output of f cycles per second will be $2\pi ft$ radians.

Thus $\qquad e = E_m \sin 2\pi ft$

or $\qquad e = E_m \sin \omega t$

where ω (Greek letter 'omega') $= 2\pi f$ radians per second.

The average and r.m.s. values of a sine wave are of importance. They are

$$\text{average value} = \frac{2 \times \text{maximum value}}{\pi} \text{ or } 0\cdot637 \times \text{maximum value}$$

$$\text{r.m.s. value} = \frac{\text{maximum value}}{\sqrt{2}} \text{ or } 0\cdot707 \times \text{maximum value}$$

$$\text{form factor} = \frac{0\cdot707 \text{ maximum}}{0\cdot637 \text{ maximum}} = 1\cdot11$$

Values for alternating systems are always given as r.m.s., unless otherwise stated.

EXAMPLE 10.5

Find the maximum and average values for a 240 V supply.

$$\text{r.m.s. value} = 0\cdot707 \times \text{maximum value}$$

Therefore

$$V_m = \frac{\text{r.m.s.}}{0\cdot707}$$

$$= \frac{240}{0\cdot707} \text{ volts}$$

$$= 339 \text{ V}$$

$$\text{Average} = 0\cdot637 \times V_m$$

$$= 0\cdot637 \times 339 \text{ volts}$$

$$= 216 \text{ V}$$

or

$$\text{Average} = \frac{\text{r.m.s.}}{\text{form factor}}$$

$$= \frac{240}{1\cdot11} \text{ volts}$$

$$= 216 \text{ V}$$

10.5 PHASOR REPRESENTATION AND PHASE DIFFERENCE

Wave diagrams, examples of which are shown in Figs. 10.1, 10.2 and 10.3, are an extremely useful way of depicting alternating values, but they are tedious to draw exactly. An alternative method of representing an alternating quantity which varies sinusoidally is a straight line called a **phasor**, its length being proportional to the value represented. The phasor is assumed to pivot at the end without an arrowhead, and to revolve anticlockwise once for every cycle of the system it represents. Consider such a phasor rotating through a complete revolution (Fig. 10.5). If the vertical height of its moving tip at various instants is transferred to a graph as shown, a sine wave results.

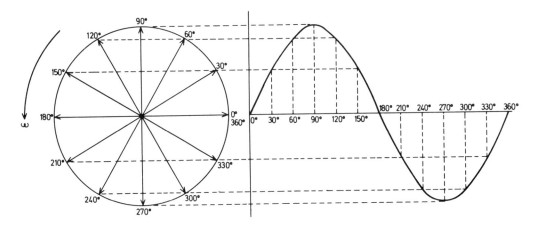

Fig. 10.5 Representation of sine wave by rotating phasor

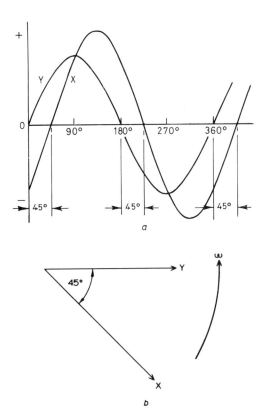

Fig. 10.6 *a* Wave diagrams of two alternating quantities, *X* lagging *Y* by 45°
b Phasor diagram for waves of *a*

Now consider two alternating quantities (Fig. 10.6a). Quantity *X* passes through zero, going positive 45° after quantity *Y*, so we say that '*X* lags *Y* by 45°', or, alternatively, '*Y* leads *X* by 45°'. The angle of 45°

between the two quantities is the **phase angle**, and, if it is unknown, it is denoted by the symbol ϕ (small Greek letter 'phi'). The phasor diagram for the arrangement is shown in Fig. 10.6b, the lengths of the phasors representing the r.m.s. values of the alternating quantities, and the angle between them being the same as the phase angle of 45°, with Y leading X, both rotating anticlockwise.

It is often necessary to add together two alternating values. If they are **in phase** (that is, if there is no displacement of phase between them), they can be added in the same way as d.c. values. Thus voltages of 100 and 150 which are in phase will sum to 250 V. If the two values are not in phase, they can be added using a wave diagram or a phasor diagram, but not by simple arithmetic.

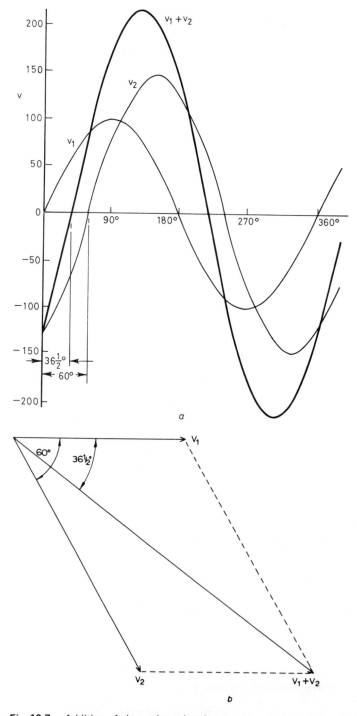

Fig. 10.7 Addition of alternating values by
 a Wave diagram
 b Phasor diagram

Fig. 10.7a gives the wave diagram of a voltage V_1 of maximum value 100 V, and a voltage V_2 of maximum value 150 V which lags V_1 by 60°. The sum of these two waves at any instant is found by adding together the instantaneous values of the individual waves for that *instant*. For example

$$\text{at } 0°, \; v_1 = 0, \quad v_2 = -130 \text{ V}, \quad \text{so } v_1 + v_2 = -130 \text{ V}$$
$$\text{at } 30°, \; v_1 = 50 \text{ V}, \quad v_2 = -75 \text{ V}, \quad \text{so } v_1 + v_2 = -25 \text{ V}$$
$$\text{at } 60°, \; v_1 = 86 \cdot 6 \text{ V}, \quad v_2 = 0, \quad \text{so } v_1 + v_2 = +86 \cdot 6 \text{ V}$$

and so on.

These two values can be added more quickly by using a phasor diagram (Fig. 10.7b). V_1 and V_2 are first drawn to scale and their **phasor sum** $V_1 + V_2$ is found by completing the parallelogram as indicated (see Section 3.5 for instructions on completing the parallelogram).

Although a line whose length and direction indicate an alternating current is now called a phasor, it was for many years referred to as a **vector**, and the terms vector and vector diagram are still commonly in use.

10.6 RESISTIVE A.C. CIRCUIT

If an alternating voltage of $v = V_m \sin \omega t$ is applied to a resistor, the instantaneous current

$$i = \frac{v}{R} = \frac{V_m \sin \omega t}{R}$$

and the current will be given by $i = I_m \sin \omega t$

Thus
$$I_m \sin \omega t = \frac{V_m}{R} \sin \omega t$$

so
$$I_m = \frac{V_m}{R}, \text{ or, by using r.m.s. values, } I = \frac{V}{R}$$

There is no phase difference between v and i, and the circuit calculations will be carried out in the same way as for a d.c. circuit. Circuit, wave and phasor diagrams are shown in Fig. 10.8.

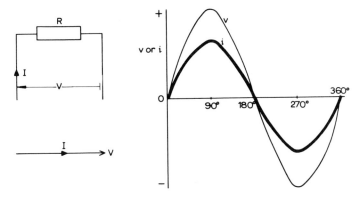

Fig. 10.8 Circuit, wave and phasor diagrams for resistive a.c. circuit

EXAMPLE 10.6

A 240 V a.c. supply is connected to an 80 Ω resistor. Calculate the resulting current flow.

$$I = \frac{V}{R}$$
$$= \frac{240}{80} \text{ amperes}$$
$$= 3 \text{ A}$$

The current and voltage used are r.m.s. values.

10.7 INDUCTIVE A.C. CIRCUIT

The property of self inductance was considered in Chapter 9, and will be more fully discussed in Volume 2. Briefly, any coil of wire which sets up a magnetic field when it carries a current has this property, so that motor windings, coils for relays, bells and telephones possess **self inductance** (symbol L).

If current changes in such a coil, an e.m.f. will be induced to oppose the change. If the current is increasing, the e.m.f. will oppose the supply voltage to limit the rate of increase, and, if decreasing, will try to keep the current flowing. The unit of self inductance is the **henry** (symbol H).

Every coil of wire must possess resistance, and, because of this resistance, it is not practicable to produce a nonresistive or 'pure' inductance. However, in this Section, we shall assume that a pure inductance exists, and examine the result of applying an alternating voltage to it.

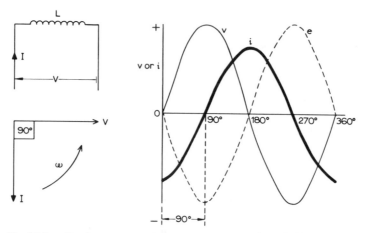

Fig. 10.9 Circuit, wave and phasor diagrams for purely inductive a.c. circuit

Circuit, wave and phasor diagrams are shown in Fig. 10.9. The e.m.f. induced in the coil must be in opposition to the applied voltage, so, on the wave diagram, v and e are drawn with a phase displacement of 180°. The induced e.m.f. depends on the rate of change of current $\{e = L(I_1 - I_2)/t\}$, so that, when e is zero, the rate of change of current must be zero; with a sinusoidally varying value, this only occurs at the maximum points, so current must be at maximum when e is zero. Induced e.m.f. must also be maximum when the rate of change of current is maximum; since this maximum occurs as the current passes through zero, maximum e.m.f. must coincide with zero current. When the current is going positive, the e.m.f. induced must oppose this change of current, and will therefore be negative. The current wave diagram can therefore be drawn, and can be seen to **lag** the applied voltage by 90°. The phasor diagram can thus be drawn as shown, induced e.m.f. being omitted.

We have assumed that the inductive circuit has no resistance, but, since the resulting current flow is not infinite, it must be limited by some property other than resistance. This property is called the **inductive reactance** of the coil (symbol X_L), and it can be shown that

$$X_L = \frac{V}{I} = 2\pi f L = \omega L$$

where
X_L = inductive reactance of the coil, Ω
V = voltage applied to the coil, V
I = resulting current flow, A
f = supply frequency, Hz
L = coil inductance, H
$\omega = 2\pi f$.

It will be noticed that, when $f = 0$, the inductive reactance will be zero. Thus the inductance of a coil has no effect on the steady flow of direct current through it, which is limited only by the coil resistance.

EXAMPLE 10.7

A coil has a self inductance of 0·318 H and negligible resistance. Calculate its inductive reactance and the resulting current if connected to (a) a 240 V, 50 Hz supply; (b) a 100 V, 400 Hz supply.

(a)
$$X_L = 2\pi f L$$
$$= 2\pi \times 50 \times 0.318 \text{ ohms}$$
$$= 100 \, \Omega$$

$$I = \frac{V}{X_L}$$
$$= \frac{240}{100} \text{ amperes}$$
$$= 2.4 \text{ A}$$

(b)
$$X_L = 2\pi f L$$
$$= 2\pi \times 400 \times 0.318 \text{ ohms}$$
$$= 800 \, \Omega$$

$$I = \frac{V}{X_L}$$
$$= \frac{100}{800} \text{ ampere}$$
$$= 0.125 \text{ A}$$

10.8 CAPACITIVE A.C. CIRCUIT

Capacitance will be treated in greater detail at the beginning of *Electrical craft principles* Vol. 2. A capacitor is a device which is capable of storing an electric charge when a potential difference is applied to it, and can be considered to consist of two conducting plates which are very close together, but are separated by an insulator called the dielectric.

The symbol for capacitance is the letter C. The unit of capacitance is the **farad** (symbol F), but this unit is too large for most practical purposes, so the millionth part of a farad, or **microfarad** (symbol μF) is used.

$$1 \mu F = 1 \times 10^{-6} \text{ farad}$$

If a direct voltage is connected to a capacitor, negatively charged electrons will be attracted from the plate connected to the positive terminal of the supply, and will be repelled from the negative terminal on to the other plate. Thus one plate will have a surplus of electrons, and the other is deficient of the same number of electrons. In this condition, the plate system is said to be **charged**, and the potential difference between the plates will increase until it equals the supply voltage, when the drift of electrons will cease.

If a capacitor is connected to a d.c. supply, the flow of current will die away quickly as the capacitor charges. Should, however, the supply be from an a.c. source, the capacitor will alternatively charge and discharge with opposite polarity. Thus, although no current actually passes through the capacitor, an alternating current exists in the circuit, which can be measured by means of a suitable ammeter.

If an alternating voltage is applied to an uncharged capacitor, as the voltage passes through zero going positive, the current will immediately reach its maximum value as the capacitor starts to charge. As the charge increases, charging current will fall, reaching zero when the voltage becomes steady, which it does for an instant at its maximum value. As the voltage falls, the capacitor will discharge, and a negative current results. This pattern repeats as shown in the wave diagram of Fig. 10.10, which also includes circuit and phasor diagrams. This shows us that, in a capacitive circuit, current *leads* supply voltage by 90° (compare with Fig. 10.9, which shows current lagging supply voltage by 90° in a purely inductive circuit).

The flow of alternating current to a capacitor is determined by its **capacitive reactance** (symbol X_c), which is measured in ohms. It can be shown that

$$X_c = \frac{V}{I} = \frac{1}{2\pi f C} = \frac{1}{\omega C}$$

where
X_c = capacitive reactance, Ω
V = supply voltage, V
f = supply frequency, Hz
I = circuit current, A
C = circuit capacitance, F
$\omega = 2\pi f$.

Since the capacitance of capacitors are more often measured in microfarads, the expression can be written as

$$X_c = \frac{10^6}{2\pi f C'}$$

where C' = circuit capacitance, μF.

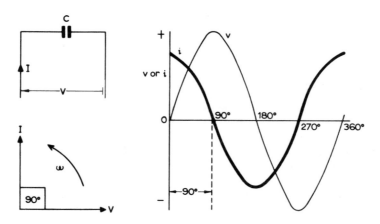

Fig. 10.10 Circuit, wave and phasor diagrams for capacitive a.c. circuit

EXAMPLE 10.8

Calculate the capacitive reactance at 50 Hz of the following capacitors: (a) 2 μF, (b) 8 μF, (c) 25 μF. Calculate also the current which will flow to each when connected to a 240 V, 50 Hz supply.

(a)
$$X_c = \frac{10^6}{2\pi f C'}$$
$$= \frac{10^6}{2\pi \times 50 \times 2} \text{ ohms}$$
$$= 1590 \, \Omega$$

$$I = \frac{V}{X_c}$$
$$= \frac{240}{1590} \text{ ampere}$$
$$= 0{\cdot}151 \text{ A}$$

(b)
$$X_c = \frac{10^6}{2\pi f C'}$$
$$= \frac{10^6}{2\pi \times 50 \times 8} \text{ ohms}$$
$$= 398 \, \Omega$$

$$I = \frac{V}{X_c}$$
$$= \frac{240}{398} \text{ ampere}$$
$$= 0.603 \text{ A}$$

(c)
$$X_c = \frac{10^6}{2\pi f C'}$$
$$= \frac{10^6}{2\pi \times 50 \times 25} \text{ ohms}$$
$$= 127 \, \Omega$$

$$I = \frac{V}{X_c}$$
$$= \frac{240}{127} \text{ amperes}$$
$$= 1.89 \text{ A}$$

10.9 TRANSFORMER

The transformer is undoubtedly the most important of all electrical machines. Transformers range in size from the miniature units used in some electronic applications to the huge power transformers used in power stations. Although the methods of construction may differ widely, all transformers follow the same basic principles. Most transformers are required to provide an output voltage which is greater or less than that applied to the transformer.

It has been shown in Chapter 9 that, if two coils are so arranged that the magnetic flux produced by one of them links with the other, a change in the current in the first coil will result in an e.m.f. being induced in the second coil. The e.m.f. induced is increased if more magnetic flux is set up for a given current by the first coil, and if more of this flux can be made to link with the second coil. For this reason, the two coils are placed on a magnetic core which forms a complete magnetic circuit.

If an alternating current flows in one coil, known as the **primary**, a resulting magnetic flux will alternate in the magnetic circuit and will link with the second coil, which is known as the **secondary**. In this way, an alternating e.m.f. will be induced in the secondary winding.

Fig. 10.11 shows the simple principle of the transformer. It should be noted that there is no electrical connection between the two windings, the link being magnetic only. Because of this, the IEE Wiring Regulations require one point on the secondary winding of most transformers to be earthed to prevent its potential from becoming large with respect to earth.

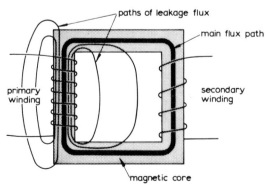

Fig. 10.11 Simplified arrangement for transformer showing main and leakage flux paths

An e.m.f. is induced in the secondary winding only while the current in the primary is changing. For this reason, transformers are ideally used on a.c. supplies. They can sometimes be used from a d.c. supply which is rapidly switched on and off so that the current is continually rising and falling. Fig. 10.12 shows the symbol used in circuit diagrams to represent a double-wound transformer.

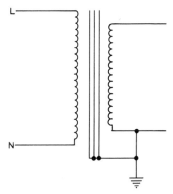

Fig. 10.12 Circuit diagram symbol for transformer

To make the transformer as efficient as possible, all the magnetic flux produced by the primary winding should link with the secondary winding. If a transformer were made exactly as shown in Fig. 10.11, a lot of the magnetic flux produced by the primary would take paths other than the path provided by the iron core.

To reduce this **leakage flux**, primary and secondary windings may be split into sections, half of each winding being placed on each side of the core. Primary and secondary may also be wound one over the other. For power transformers, the core-type construction shown in Fig. 10.13 is often chosen. The windings are placed on the centre limb, the side limbs providing a path for return flux.

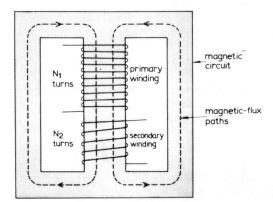

Fig. 10.13 Practical arrangement for simple transformer

The e.m.f. induced in each turn of the transformer secondary will depend on the rate of change of the magnetic flux which links with it. Since the turns of a winding are all in series with each other, the total e.m.f. induced in a winding will be the product of the e.m.f. per turn and the number of turns.

If leakage is neglected, the same changing magnetic flux links both primary and secondary windings, so the e.m.f. per turn will be the same for both windings. The total induced e.m.f. is therefore proportional to the number of turns, and the ratio of e.m.f.s in the two windings will be the same as the ratio of their numbers of turns; that is

$$\frac{E_1}{E_2} = \frac{N_1}{N_2}$$

where
E_1 = e.m.f. induced in the primary, V
E_2 = e.m.f. induced in the secondary, V
N_1 = number of primary turns
N_2 = number of secondary turns.

In practice, both windings will have resistance, so that the voltages across the coils on load will not be quite the same as the e.m.f.s induced in them, the differences being the voltage drops in the windings. Power-transformer windings have low resistance, and only a small error will be introduced by assuming that the terminal voltage is the same as the induced e.m.f., so that

$$E_1 = V_1 \quad \text{and} \quad E_2 = V_2$$

where
V_1 = voltage applied to the primary
V_2 = terminal voltage of the secondary.

Therefore
$$\frac{V_1}{V_2} = \frac{N_1}{N_2}$$

or   voltage ratio = turns ratio

EXAMPLE 10.9

A transformer is wound with 480 turns on the primary and 24 turns on the secondary. What will be the secondary voltage if the primary is fed at 240 V?

$$\frac{V_1}{V_2} = \frac{N_1}{N_2} \quad \text{so} \quad V_2 = V_1 \frac{N_2}{N_1}$$

$$V_2 = 240 \times \frac{24}{480} \text{ volts}$$

$$= 12 \text{ V}$$

EXAMPLE 10.10

A transformer with a turns ratio of 2:9 is fed from a 240 V supply. What is its output voltage?

Transformer ratios are always given in the form 'primary : secondary'.
This transformer has two turns on the primary for every nine on the secondary, so $N_1/N_2 = 2/9$.

$$V_2 = V_1 \frac{N_2}{N_1}$$

$$= 240 \times \frac{9}{2} \text{ volts}$$

$$= 1080 \text{ V}$$

If the secondary voltage is greater than the primary, the transformer is referred to as **stepup**, and, if smaller, as **stepdown**.

If we neglect losses, the input and output powers for a transformer will be the same. Put simply,

$$V_1 I_1 = V_2 I_2$$

where I_1 and I_2 are primary and secondary currents, respectively. Transposing, and adding the turns ratio to the expression, we have

$$\frac{V_1}{V_2} = \frac{I_2}{I_1} = \frac{N_1}{N_2}$$

Thus, if a transformer steps up the voltage from primary to secondary, the current will be less in the secondary than in the primary.

EXAMPLE 10.11

A transformer has a turns ratio of 5:1, and is supplied at 240 V when the primary current is 2 A. Calculate the secondary current and voltage

$$V_2 = \frac{V_1 N_2}{N_1}$$

$$= \frac{240 \times 1}{5} \text{ volts}$$

$$= 48 \text{ V}$$

$$I_2 = \frac{V_1 I_1}{V_2}$$

$$= \frac{240 \times 2}{48} \text{ amperes}$$

$$= 10 \text{ A}$$

If an iron core has an alternating magnetic flux set up in it, it will get hot owing to what are called 'core losses'. These losses are subdivided into

(a) **hysteresis losses:** This subject will be considered more fully in *Electrical craft principles* Vol. 2. Hysteresis losses can never be removed, but can be reduced by making the magnetic core of an iron containing a small percentage of silicon.

(b) **eddy-current losses:** The iron core is itself a conductor, and will have an e.m.f. induced in it by the changing magnetic flux which it carries. The e.m.f. causes a current to eddy back and forth in the iron, and this current causes heating.

To reduce eddy currents to a minimum, the core of a transformer is subdivided into small parts; each part has only a small share of the total induced e.m.f., and has high resistance owing to its small cross-section, so that eddy currents are small.

In power transformers, this subdivision is carried out by building up the core with layers of thin plates, called **laminations.** Each lamination is insulated from its neighbours by being varnished or oxidised or covered with paper on one side. The laminations must be arranged so that the magnetic flux set up is along them, and not across the insulation between them, which is nonmagnetic and would reduce the strength of the field (Fig. 10.14). A coil-winding machine is used to make the transformer coils, which are then slipped

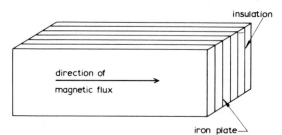

Fig. 10.14 Part of magnetic circuit built up of laminations
Lamination thickness has been exaggerated

over the core. Thus a core of the type shown in Fig. 10.13 would be made up of E and I sections, or of U and T sections. Joints in the core must fit as well as possible to allow the maximum magnetic flux to be set up; laminations are interleaved at the corners to improve the effective fit.

10.10 SUMMARY OF FORMULAS FOR CHAPTER 10

For sinusoidal systems:

$$\text{Mean (average) value} = \frac{2}{\pi} \times \text{maximum value}$$

$$= 0 \cdot 637 \times \text{maximum value}$$

$$\text{Effective (r.m.s.) value} = \frac{1}{\sqrt{2}} \times \text{maximum value}$$

$$= 0 \cdot 707 \times \text{maximum value}$$

$$\text{Form factor for any waveshape} = \frac{\text{effective value}}{\text{mean value}}$$

For a purely resistive a.c. circuit,

$$I = \frac{V}{R} \qquad R = \frac{V}{I} \qquad V = IR$$

where
V = applied voltage (effective value), V
I = resulting current (effective value), A
R = circuit resistance, Ω.

For a purely inductive a.c. circuit,

$$I = \frac{V}{X_L} \qquad X_L = \frac{V}{I} \qquad V = IX_L$$

where X_L = inductive reactance, Ω.

$$X_L = 2\pi fL \qquad f = \frac{X_L}{2\pi L} \qquad L = \frac{X_L}{2\pi f}$$

where
f = supply frequency, Hz
L = circuit inductance, H

For a purely capacitive a.c. circuit

$$I = \frac{V}{X_c} \qquad X_c = \frac{V}{I} \qquad V = IX_c$$

where X_c = capacitive reactance, Ω.

$$X_c = \frac{1}{2\pi fC} \qquad f = \frac{1}{2\pi CX_c} \qquad C = \frac{1}{2\pi fX_c}$$

where C = circuit capacitance, F

or

$$X_c = \frac{10^6}{2\pi fC'}$$

where C' = circuit capacitance, μF.

For a simple transformer, neglecting leakage,

$$\frac{V_1}{V_2} = \frac{N_1}{N_2}$$

$$V_1 = \frac{V_2 N_1}{N_2} \qquad V_2 = \frac{V_1 N_2}{N_1} \qquad N_1 = \frac{V_1 N_2}{V_2} \qquad N_2 = \frac{N_1 V_2}{V_1}$$

where
V_1 = primary voltage, V
V_2 = secondary voltage, V
N_1 = number of primary turns
N_2 = number of secondary turns

$$\frac{V_1}{V_2} = \frac{I_2}{I_1}$$

$$V_1 = \frac{V_2 I_2}{I_1} \qquad V_2 = \frac{V_1 I_1}{I_2} \qquad I_1 = \frac{V_2 I_2}{V_1} \qquad I_2 = \frac{V_1 I_1}{V_2}$$

where
I_1 = primary current, A
I_2 = secondary current, A.

10.11 EXERCISES

1. An alternating voltage completes one cycle in 1 ms. What is its frequency?

2. Explain the meaning of the term 'periodic time' as applied to an a.c. system. What is the periodic time of a system with a frequency of 60 Hz?

3 An alternating voltage is triangular in shape, rising at a constant rate to a maximum of 300 V in 0·01 s, and then falling to zero at a constant rate in 0·005 s. The negative halfcycle is identical in shape to the positive halfcycle. Calculate
 (a) the supply frequency
 (b) the average voltage
 (c) the r.m.s. voltage
 (d) the form factor.

4 A voltage is represented by a sine wave, and has a maximum of 200 V. Calculate its r.m.s. and average or mean value. (NCTEC)

5 Explain, with the aid of a diagram, the meaning of the following terms as applied to an alternating e.m.f.:
 (a) maximum value,
 (b) root-mean-square value,
 (c) cycle.
 If the frequency of the supply is 50 Hz, what is the time taken for the completion of one cycle? (ULCI)

6 The positive halfcycle of an alternating current has the following values at 1 ms intervals:

t (ms)	0	1	2	3	4	5	6	7	8	9
I (A)	0	7	13	19	24·5	30·5	35·5	40	44·5	48
t (ms)	10	11	12	13	14	15	16	17	18	19	20
I (A)	50·5	52	52·5	52·5	51·5	49·5	45·5	39·5	31	15	0

Plot this curve, and hence calculate
 (a) the supply frequency
 (b) the mean value of the current
 (c) the effective value of the current
 (d) the form factor
 (e) the instantaneous value of current after 27 ms.

7 Using sine tables, draw out the positive halfcycle of a sinusoidal voltage of peak value 200 V, and calculate the average voltage, the r.m.s. voltage and the form factor.

8 A sinusoidal current has an effective value of 10 A. Calculate its average and maximum values.

9 Calculate the mean and peak voltages of a 240 V sinusoidal supply.

10 Explain the following terms applied to an alternating-current wave: (a) maximum value, (b) average value, (c) r.m.s. value.
 If the maximum value of a sine wave is 300 A, give the average and r.m.s. values. (C & G)

11 (a) Draw *freehand* two complete cycles of a sinusoidal alternating current of peak value 100 A. Label (i) one point of current zero, (ii) one complete cycle, (iii) one negative peak.
 (b) What is the r.m.s. value of this sine wave?
 (c) Why is the *r.m.s.* value (*effective* value) normally used to specify the current value? (C & G)

12 Draw freehand two complete cycles of a sinusoidal alternating current. Label one negative peak, one positive peak, one current zero and one positive halfcycle.
 Why is an alternating current usually specified by its r.m.s., or effective, value?
 If the peak value of a sine wave of current were 200 A, what would be its r.m.s. value? (C & G)

13 An alternating current generator has four poles. A conductor on the rotating armature generates a sine wave of e.m.f. of maximum value 10 V.
 (a) Draw a graph baseline to represent one revolution, mark in the pole-centre positions, and then sketch in the e.m.f. wave.
 (b) What is the r.m.s. value?
 (c) How many cycles are produced per revolution? (C & G)

14 Two sinusoidal currents, each of 100 A peak value, are added. One starts at zero time, and the other a quarter cycle later. Draw the two waves, and the wave of their sum, and measure
 (a) the peak value of the sum
 (b) the degrees after zero time at which the sum reaches its first peak.
 The following values may be used:

θ	0°	30°	60°	90°	120°	150°	180°
sin θ	0	0·5	0·866	1	0·866	0·5	0

(C & G)

15 Draw a scale phasor diagram to show a current of 15 A leading a current of 10 A by 30°. Find the resultant of these phasors, and its phase relative to the 10 A current.

16 Find the resultant of two 100 V a.c. supplies which are connected in series but 120° out of phase (a) by phasor addition, (b) by wave addition.
Find the phase of the resultant relative to either 100 V supply.

17 Calculate the current in a 20 Ω resistor connected to a supply at
(a) 240 V, 50 Hz;   (b) 115 V, 400 Hz;   (c) 1000 V, 100 Hz;   (d) 200 V, 30 Hz;   (e) 300 V, 60 Hz.

18 Calculate the inductive reactance of the given nonresistive inductor, and the current when connected to the given supply:
(a) 1 H connected to a 240 V, 50 Hz supply
(b) 20 mH connected to a 415 V, 60 Hz supply
(c) 0·15 H connected to a 100 V, 400 Hz supply.

19 Calculate the capacitive reactance of the given capacitor and the current when connected to the given supply.
(a) 10 μF connected to a 240 V, 50 Hz supply
(b) 20 nF connected to a 12 V, 1 kHz supply
(c) 300 μF connected to a 3 V, 30 Hz supply
(d) 1 pF connected to a 24 V, 50 kHz supply
(e) 40 nF connected to a 1 V, 1 Mhz supply.

20 Make a labelled sketch of a square-cross-section core for a single-phase transformer showing its detailed construction. Describe briefly its various components and the precautions to be taken during its assembly. (C & G)

21 A transformer connected to a 240 V supply has 600 primary turns, and is required to have a secondary voltage of 40 V. How many turns must be wound on the secondary?

22 A transformer with a turns ratio of 500:40 has an output of 10 V. What voltage is applied to the primary winding?

23 A transformer primary winding connected across a 415 V supply has 600 turns. How many turns must be wound as the secondary if an output of 1328 V is required?

24 Make sketches indicating current flow and resulting magnetic field when current flows in a solenoid with a steel core in the shape of a closed ring. Explain how this construction can form the basis of a transformer if supplied with alternating current. (C & G)

25 Draw a sketch of a transformer showing core and coils and mark in the flux path. Explain how the transformer works. (C & G)

26 A transformer is supplied at 240 V. The primary winding has 4800 turns, and takes 1 A for every 10 A delivered by the secondary. Calculate (a) the number of turns on the secondary, (b) the secondary voltage. (C & G)

27 Describe, with the aid of a sketch, a simple stepup transformer and explain its action. (ULCI)

28 A transformer has primary and secondary voltages of 720 V and 300 V, respectively. What will be the primary current when the secondary delivers 10 A?

29 A transformer has 1500 primary turns and 75 secondary turns. What current will the secondary provide if the primary carries 5 A?

30 A transformer has 200 primary turns, and is fed at 120 V. If primary and secondary currents are measured at 1·5 A and 120 mA, respectively, calculate the number of secondary turns and the secondary voltage.

Chapter 11

Electric motor principle

11.1 INTRODUCTION

In Chapter 9, it was explained that an electric current gives rise to a magnetic field. If a second current-carrying conductor is placed in such a field, it is subjected to an electromagnetic force. This force can be used to drive the conductor, and an electric motor results.

It is interesting to reflect how much our civilisation depends on the electromagnetic principles of the generator and the motor. Without them, the world would be a very different place.

11.2 FORCE ON A CURRENT-CARRYING CONDUCTOR LYING IN A MAGNETIC FIELD

If a conductor lies at right angles to a magnetic field, it experiences a force when a current passes through it. We can verify this natural law by using a piece of wire, a permanent magnet and a battery. If the wire is placed between the magnet poles and its ends momentarily connected across the battery terminals, it will jump from its position.

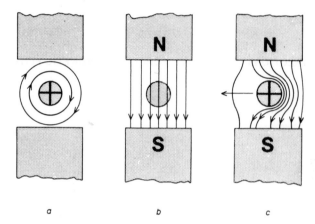

Fig. 11.1 Force on current-carrying conductor lying in magnetic field

Fig. 11.1 explains why this force occurs. Fig. 11.1a shows the magnetic field due to a conductor, drawn in cross-section, which is carrying a current into the plane of the paper. Fig. 11.1b shows the magnetic field due to the magnetic poles, between which the conductor is situated, when the conductor carries no current. Since lines of magnetic flux never cross, the two fields cannot exist simultaneously in their individual forms, and the resultant field takes up the shape shown in Fig. 11.1c. The stronger field to the right of the conductor tries to contract, and exerts a force on the conductor in much the same way as if it were a stone in a catapult. If free to do so, the conductor will move to the left. Should the conductor be moved out of the influence of the magnetic field due to the poles, it will cease to have a force applied to it.

If the reader redraws the poles, conductor and magnetic field, he will find that, if either the polarity of the magnet or the current in the conductor is reversed, the force on the conductor will be reversed. If both are reversed, the force remains in the same direction. Clearly, it is important to be able to calculate the force on the conductor in given circumstances.

Experiment shows that, provided the conductor is at right angles to the field,

$$F = BIl$$

where
F = force on conductor, N
B = flux density of magnetic field, T
l = length of conductor in field, m
I = current flowing in conductor, A.

EXAMPLE 11.1

A conductor, 0·2 m long, carries a current of 25 A at right angles to a magnetic field of flux density 1·2 T. Calculate the force exerted on the conductor.

$$F = Bll$$
$$= 1\cdot2 \times 0\cdot2 \times 25 \text{ newtons}$$
$$= 6 \text{ N}$$

EXAMPLE 11.2

How much current must a conductor of an electric motor carry if it is 900 mm long and is situated at right angles to a magnetic field of flux density 0·8 T, if it has a force of 144 N exerted on it?

$$F = Bll$$

Therefore
$$I = \frac{F}{Bl}$$
$$= \frac{144}{0\cdot8 \times 0\cdot9} \text{ amperes}$$
$$= 200 \text{ A}$$

11.3 RELATIVE DIRECTIONS OF CURRENT, FORCE AND MAGNETIC FLUX

It is often important to know the direction of the force on a conductor when it carries a current of given direction in a field of given polarity. One method is to draw out the field as shown above, but there is a rule which links the directions of the current, field and force and enables us to find the third if the directions of the other two are known.

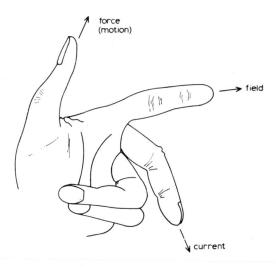

Fig. 11.2 Position of left hand for application of Fleming's left-hand rule

This is **Fleming's left-hand (motor) rule**. The thumb, the first finger and the second finger of the left hand are extended, so that all three are at right angles to each other (Fig. 11.2). If the first finger points in the direction taken by the magnetic field, and the second finger in the direction of current flow, the thumb gives the direction of motion of the conductor as a result of the force applied to it. This is easily remembered by noticing that the **F**irst finger gives the magnetic **F**ield direction, the se**C**ond finger gives the **C**urrent direction and the thu**M**b gives the direction of conductor **M**otion as a result of the force.

A little practice will show how easy this rule is to apply, but it must be carried out using the left hand, and applies only to the motor effect.

EXAMPLE 11.3

Refer to Fig. 11.3, and give
    (a)   the direction of the force on the conductor in Fig. 11.3a

(b) the polarity of the field system in Fig. 11.3b
(c) the direction of the current in Fig. 11.3c
(d) the direction of the force on the conductor in Fig. 11.3d.

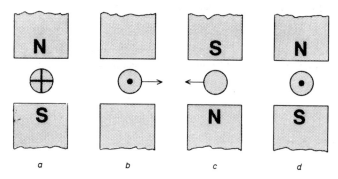

Fig. 11.3 Figures for example 11.3

Applying Fleming's left-hand rule, or sketching the magnetic-field shapes, gives the results
(a) right to left
(b) north pole at the top
(c) out of the paper
(d) left to right.

11.4 LENZ'S LAW

Consider a conductor being moved by an external source of energy at right angles to a magnetic field (Fig. 11.4). The application of Fleming's right-hand rule will show that the direction of induced e.m.f. is such as to cause a current into the plane of the paper, if the conductor forms part of a closed circuit. We now have a current-carrying conductor situated in a magnetic field. As Fleming's left-hand rule will show, the direction of the force on this conductor will be opposite to the conductor movement.

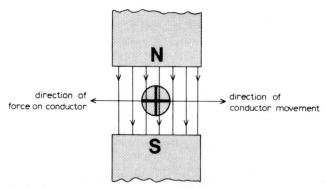

Fig. 11.4 Differing directions of conductor movement and force (Lenz's law)

Thus, movement of a conductor in a magnetic field gives rise to current in such a direction as to oppose the movement causing it. Lenz's law states that the direction of an induced e.m.f. is always such that it tends to cause a current which opposes the change inducing the e.m.f. In the case shown, extra work must be done to overcome the reverse force; the work needed to overcome it will increase as the current increases, so that, for a generator, we have to put more mechanical energy in to get more electrical energy out.

The reverse force will not, of course, completely stop the conductor. If it did so, the induced e.m.f., and hence the current producing the reverse force, would disappear.

The law also affects induction in a circuit, which is due to a change in linking magnetic flux (Chapter 9). The e.m.f. induced by the changing current will always be in such a direction as to resist that change. If the current is reducing, the e.m.f. will be in the same direction as the current, and will try to maintain it; if the current is increasing, the e.m.f. will oppose it and try to prevent the increase.

11.5 DIRECT-CURRENT-MOTOR PRINCIPLES

The direct-current motor is basically the same as the direct-current generator, which was considered in Section 9.5. Both machines are energy convertors. The generator is supplied which mechanical energy, and gives out most of this energy in electrical form. The motor takes in electrical energy and provides mechanical work.

Consider the simple rectangular-loop system shown in Fig. 11.5. This is the same as the generator arrangement of Fig. 9.7a, but, instead of providing electricity, it must be supplied with electricity, so a d.c. supply is connected to the brushes.

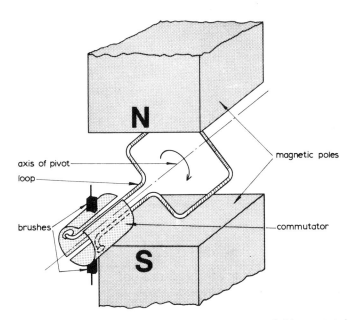

Fig. 11.5 Loop connected to simple commutator and able to rotate in magnetic field

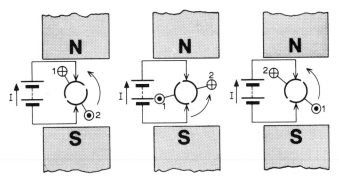

Fig. 11.6 Principle of d.c. motor

Fig. 11.6 shows the directions of the forces experienced by the conductors, which can be verified by application of Fleming's left-hand rule. The commutator reverses the current flow in a conductor at it passes from one pole to the next, so that the current in either conductor will always be the same as it passes a given pole. The direction of the force will therefore always be the same, and the loop will rotate continuously in a given direction. At the instant when the brushes are passing over the joints in the commutator, the conductors will be moving along the lines of magnetic flux, and will experience no force. In practice, the speed of rotation of the loop will keep it moving until this 'dead spot' is passed. Like the generator, the practical d.c. motor has many loops and a multisegment commutator. As a result, the force on the machine is nearly constant, and no 'dead spot' occurs.

The construction of the d.c. machine was described briefly in Section 9.5.

11.6 MOVING-COIL INSTRUMENT

In Section 11.2 we saw that the force exerted on a conductor carrying current in a magnetic field depends on the magnetic-field strength, the length of conductor in the field, and the conductor current ($F = BIl$). If field strength is made constant by the use of a permanent magnet, and the conductor is in the form of a coil of fixed length, the force must depend only on the current. Thus an instrument can be made to measure the current it carries, giving a deflection depending on the force exerted on its coil, and hence on its current.

This instrument is called the **permanent-magnet moving-coil** instrument, and is used widely for current and voltage measurements. It is rather like a miniature d.c. motor, but, instead of having a commutator to allow continuous rotation, current is fed into the coil through hair springs which also serve to limit the angle of rotation.

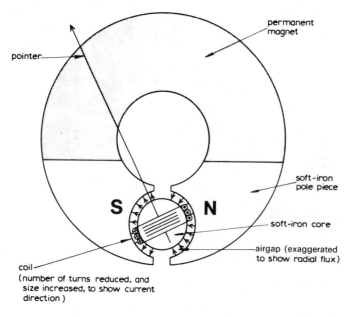

Fig. 11.7 General arrangement of permanent-magnet moving-coil instrument

In many types, the shape of the permanent magnet is similar to that shown in Fig. 11.7, the magnetic circuit being completed with shaped soft-iron pole shoes and cylindrical core, so that a radial and uniform magnetic field is set up in the airgap. The coil of fine insulated wire wound on an aluminium former is pivoted to swing in this field, and will always cut it at right angles (Fig. 11.8). Two phosphor-bronze hairsprings serve to make electrical connections to the moving coil, as well as limiting the coil swing (controlling torque) and returning the movement to the zero position when no current flows (restoring torque). The coil moves

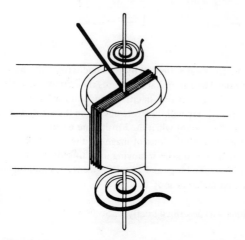

Fig. 11.8 View of permanent-magnet moving-coil movement, showing coil, pointer, spindle and springs

because, when current flows in it, it becomes a series of current-carrying conductors lying in a magnetic field. The force on the two sides of the coil turn it against the torque of the control springs, equilibrium occurring when the deflecting torque due to the current is equal and opposite to the controlling torque due to the hairsprings. A light aluminium pointer is fixed to the coil, and moves over a scale to measure the current.

With most instruments, some provision must be made to ensure that the movement comes to rest quickly at its reading, without excessive oscillation. This provision is known as **damping**. With the permanent-magnet moving-coil instrument, damping is achieved automatically, since the e.m.f. induced in the aluminium-coil former as it swings in the magnetic field causes an eddy current to flow which, obeying Lenz's law, produces a damping force which opposes the movement producing it.

The wire of which the moving coil is made is very fine so as to reduce the weight of the moving system, and cannot carry much current. In consequence, the torque on the coil is not very great, so the whole moving system is delicately mounted on jewelled bearings, and all possible precautions are taken to cut down friction. The moving system is carefully balanced, often having balance arms with adjustable weights for this purpose. Some instruments have a taut wire suspension, which replaces the spindle, bearings and springs, making the movement frictionless.

In their standard form, these instruments are limited to an angular movement of about 120°, but special movements are available to give **circular scales**, the needle being capable of swinging through an angle of about 300°. This increased scale length gives an improvement in the accuracy of reading, or alternatively, the same scale length can be accommodated on an instrument taking up less space.

Since the torque, and hence the instrument deflection, is proportional to current, the scale is linear: that is, there are equal spaces between equal divisions. Other advantages of this instrument are its accuracy, its sensitivity to small currents, and the ease with which it can be adapted for almost any value of current or voltage. The main disadvantage of the instrument is its inability to read values of alternating current at power frequencies. These currents give rise to an alternating torque; the movement, being unable to adjust to these rapid variations, remains at the zero position. Other disadvantages are that the delicate moving system is easily damaged by rough handling, and that the fine coil will not withstand prolonged overloading.

11.7 SUMMARY OF FORMULAS FOR CHAPTER 11

For a conductor lying at right angles to a magnetic field,

$$F = BIl \qquad B = \frac{F}{Il} \qquad l = \frac{F}{BI} \qquad I = \frac{F}{Bl}$$

where
F = force on conductor, N
B = flux density of magnetic field, T
l = length of conductor in magnetic field, m
I = current carried by conductor, A

11.8 EXERCISES

1. A force of 10 N is exerted on a conductor 1·5 m long when carrying a current and lying at right angles to a magnetic field of flux density 1·5 T. What is the current?

2. If a conductor lying at right angles to a magnetic field of flux density 0·12 T experiences a force of 8 N when carrying a current of 5 A, what is the effective length of the conductor in the magnetic field?

3. 20 mm of a conductor carrying a current of 15 mA is situated at right angles to a magnetic field, and experiences a force of 0·33 mN. What is the field flux density?

4. With the aid of a sketch, show how a force is produced on a current-carrying conductor lying in a magnetic field. Directions of the current, magnetic field and force should be shown. (NCTEC)

5. Fig. 11.9 gives examples of a current-carrying conductor lying in a magnetic field. State
    (a) the direction of the force on the conductor in Fig. 11.9a;
    (b) the direction of the force on the conductor in Fig. 11.9b;
    (c) the direction of the current in Fig. 11.9c;
    (d) the polarity of the magnet in Fig. 11.9d.

6 What force is experienced by a busbar 2 m long, which, under fault conditions, carries a current of 20 000 A in a magnetic field of flux density 100 mT?

7 One conductor on the coil of a moving-coil instrument is 10 mm long, and experiences a force of 2 μN when carrying a certain current. The airgap flux density is 0·8 T. What is the coil current?

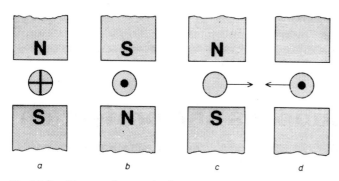

Fig. 11.9   Diagrams for exercise 5

8 A moving-coil loudspeaker is required to provide a force of 0·2 N when its coil, which has an effective conductor length of 15 m, carries a current of 12 mA. What flux density must be set up by the magnet?

9 What length of conductor must be present in the armature of a motor which has a pole flux density of 0·6 T and is required to provide a turning force of 300 N when the conductor current is 25 A?

10 (a) Consider a single conductor carrying direct current and lying in the magnetic field between the poles of a 2-pole d.c. motor.
Make a sketch or diagram illustrating the above. Assuming your own directions of magnetic field and current, indicate clearly the direction in which the conductor will tend to move.
(b) A conductor 0·3 m long lies at right angles to a magnetic field of intensity 1·6 T and carries a current of 25 A. Calculate the force on the conductor. (C & G)

11 Sketch a 2-pole d.c. motor showing armature, poles, commutator and brushes, and explain how a continuous torque in the same direction is obtained. (C & G)

12 Sketch a moving-coil instrument and label the main parts. State two disadvantages of this type of instrument. (ULCI)

13 Explain why a moving-coil instrument is unsuitable for use in a.c. circuits. (NCTEC)

Chapter 12

Practical supplies and protection

12.1 INTRODUCTION

The purpose of this chapter is to show how the basic theory already considered is put into practice to provide supplies and systems.

There are a number of basic types of supply, and these will be considered in greater detail in the succeeding sections. Electricity is dangerous, as its misuse can generate heat, which results in fire. If the human body becomes part of an electric circuit, the resulting 'shock' can result in burns, or may be fatal. This chapter will begin to show the principles to be followed if these dangers are to be avoided.

12.2 DIRECT-CURRENT SUPPLIES

Direct-current supplies are not available from supply authorities. The reasons for this are listed in Section 10.2, which gives some of the more important advantages of a.c. systems.

Although mains supplies are all of the alternating-current type, it must be remembered that d.c. supplies are still in very wide use. For example, where an emergency supply is needed, it is often most economical to provide a battery of secondary cells. These cells give a d.c. supply, usually for emergency lighting in the event of a mains failure. Again, the electrical systems of motor vehicles are almost always of the d.c. type, so that batteries can be used for engine starting and other services (radio, parking lights etc.) which are required when the engine is not running. The modern practice is to fit an a.c. generator (alternator) to motor vehicles, but the output is converted to a d.c. system before use.

Equipment intended to break current on d.c. supplies is always of heavier construction (and thus more expensive) than its a.c. counterpart. This is because the voltage across a switch, or a similar break in a circuit, is continuous, and tends to maintain any arc that may have formed as the contacts separated. Such an arc dissipates a great deal of heat, and must be broken as soon as possible. Wide separation of switch contacts is thus required. More complicated methods of breaking the arc, such as the provision of arc chutes, or immersion of contacts in oil, are necessary in heavy-current systems.

12.3 SINGLE-PHASE A.C. SUPPLIES

Most consumers are fed by means of a single-phase a.c. supply. Two wires are used, one called the **phase conductor** and usually coloured red, and the other is called the **neutral conductor** and is coloured black. Both phase and neutral conductors are called 'live' because both normally carry current. The neutral is usually earthed, the earth wire being coloured green and yellow (Section 12.5). For flexible cords, the phase conductor is coloured brown, the neutral conductor blue, and the earth conductor green and yellow.

Most houses have a single-phase a.c. supply, which is fed in by means of a 2-core cable when the supply is underground, or two separate overhead conductors. The majority of single-phase supplies are obtained by connection to a 3-phase supply.

12.4 3-PHASE A.C. SUPPLIES

The standard method of transmitting and distributing electrical energy in most countries is by use of a 3-phase a.c. system.

If a simple wire loop is rotated in a uniform magnetic field, the e.m.f. induced in it varies sinusoidally (Chapter 9). If three such loops are fixed together so that each makes an angle of 120° with the other two, we have an elementary system which will provide a symmetrical 3-phase supply if rotated in a uniform magnetic field. A section through such a system is shown in Fig. 12.1, the coils being identified by the colours red (R), yellow (Y) and blue (B). Each coil will have two ends. If the ends R_1, Y_1 and B_1 are connected together, and the ends R, Y and B are brought out to the slip rings, the coils being rotated in a clockwise direction, the outputs of the coils will be as shown in the wave diagram of Fig. 12.2. The phasor diagram for the induced e.m.f.s is also shown. It will be seen that the electrical-phase displacement between the e.m.f.s is 120°, the same as the physical displacement of the coils.

There are a number of reasons for the adoption of 3-phase transmission and distribution systems. Some of the most important are

Practical supplies and protection   155

(a) less copper (or aluminium) is needed for the conductors of a 3-phase system which transmits a given power at a given voltage over a given distance than for a similar single-phase system
(b) 3-phase motors have many advantages over single-phase motors, including smaller size, steady torque output, and the ability to self-start
(c) when connected in parallel, single-phase generators present difficulties which do not occur with 3-phase generators.

Three wires, one connected to each of the slip rings of the wire loops of Fig. 12.1, are necessary to carry a 3-phase supply. These wires are called the **lines** and are identified by the colours red, yellow and blue. The currents flowing in the lines are called **line currents** (symbol I_L) and the p.d. between any two lines is called the **line voltage** (symbol V_L). A fourth wire, called the **neutral**, coloured black, and usually at earth potential, is often used in conjunction with a 3-phase supply. Three separate loads can be connected to a three-phase

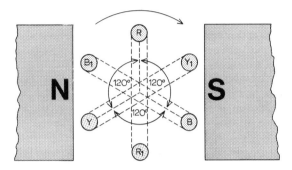

Fig. 12.1   Arrangement of simple wire-loop generator to produce 3-phase supply

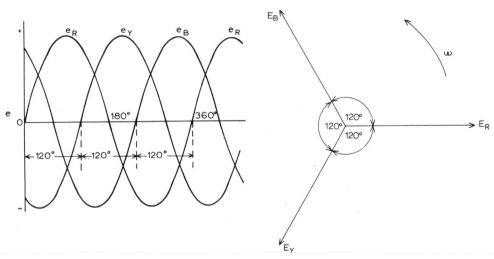

Fig. 12.2   Wave and phasor diagrams for 3-phase supply

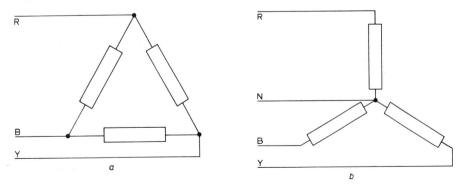

Fig. 12.3   Balanced loads connected to 3-phase supply
   a Delta or mesh connection   *b* Star connection. Fourth (neutral) wire may be omitted if load is balanced

supply in either of two ways, the current in each load being known as the **phase current** (symbol I_P), and the p.d. across each load as the **phase voltage** (symbol V_P). The methods of connecting to a 3-phase supply are known as the **delta (or mesh)** and **star** connections (Fig. 12.3).

For the delta connection, line and phase voltages are the same, but line current is greater than phase current. For balanced systems,

$$V_L = V_P$$
$$I_L = \sqrt{3}\, I_P$$
delta connection

For the star connection, line and phase currents are the same, but line voltage is greater than phase voltage. For balanced systems,

$$V_L = \sqrt{3}\, V_P$$
$$I_L = I_P$$
star connection

These relationships will be proved in Volume 2.

Three-phase supplies are taken into large industrial and commercial premises, and are used directly to supply heavy individual loads, such as large motors, heaters, furnaces etc. If the load is unbalanced, that is if each of the three parts of the load takes a different current, a star connection with a neutral (known as a 3-phase 4-wire supply) is usually used. If the load is balanced, no neutral is needed, and the load may be star- or delta-connected.

For most domestic and many commercial and industrial applications, the higher voltage of the 3-phase system (415 V, which is $\sqrt{3}$ × 240 V, between lines is standard) is not required, and single-phase supplies are taken from the 3-phase 4-wire supply. Each single-phase supply is connected to one phase and the neutral, so that three separate 2-wire single-phase supplies can be obtained from a 4-wire 3-phase supply (Fig. 12.4). These 240 V supplies are used for lighting and general small-load purposes, such as cleaning, heating, ventilating etc. Some single-phase welders operate at 415 V, and the diagram also shows how such a connection can be made.

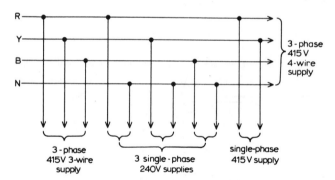

Fig. 12.4 Supplies obtainable from standard 3-phase 4-wire supply

The standard colours for the cables of a 3-phase system are red, yellow and blue, with black for the neutral. For 3-phase flexible cords, all 3-phase conductors are brown, but numbered 1, 2 and 3, respectively. The neutral is then blue.

Care is taken in designing an electrical installation to 'balance the load'. To do this, the loads on the three separate phases are kept as nearly equal as possible, always bearing in mind that it may be dangerous for the areas served by different phases to overlap. This is because 415 V exists between single-phase circuits derived from different phases of a 3-phase supply.

12.5 EARTHING

The general mass of earth is made up almost entirely of materials which are reasonable electrical conductors themselves, or are made so by being moist. From this, it follows that a current will flow to earth through a conductor which connects a live system to earth, provided some other point of the system at a different potential is also connected to earth.

In practice, the neutral at the supply mains is almost always connected to the general mass of earth, by connecting a conductor from the neutral to a metal pipe or rod driven into the ground, or to metal plates or tapes buried in it, and called an **earth electrode**. The complete circuit taken by the current to earth, or **earth fault current**, in the event of an accidental connection, or fault, to earth is shown in Fig. 12.5. This is called the **earth fault loop**.

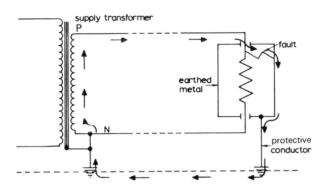

Fig. 12.5 Earth fault loop. The path of fault current in the loop is shown by heavy arrows

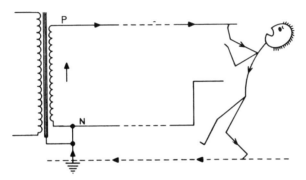

Fig. 12.6 Electric shock due to contact with phase conductor and earth (direct contact)

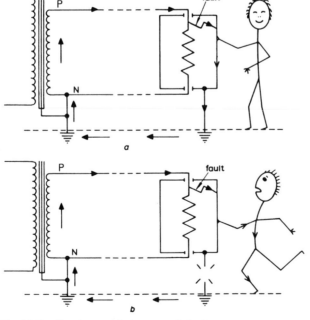

Fig. 12.7 Shock to earth due to earth fault
  a Protective conductor is intact, and short-circuits path through body
  b Broken protrective conductor results in shock (indirect contact)

If a human body forms part of this loop, by touching phase conductor while also in contact with earth, the person concerned will receive a shock. The circuit shown in Fig. 12.6. This possiblity of shock is the disadvantage of earthing, but is comparatively easy to prevent by insulating and protecting live conductors.

One advantage of earthing is that if all the exposed metalwork of an installation is connected to earth, no significant potential difference should exist for very long between that metalwork and earth. This means that there can be no possibility of a serious shock being received from such metalwork, which includes metal conduits, cable sheaths, motor casings, metal fire reflectors and so on. The exposed metalwork is connected together and to earth, by means of the **protective conductor**. This consists of the metal cable sheath, steel cable trunking or conduit, a separate earth wire, or any combination of these items. The protective conductor is connected to earth with a conductor called the **earthing conductor**, which is attached to the incoming cable sheath or an earth electrode. It is vitally important to ensure that the earth continuity conductor remains intact and is always connected to earth. Fig. 12.7a shows how a continuous earthing path 'shorts out' the shock path through the human body, and Fig. 12.7b shows how the body may become part of the circuit if the earth path is broken.

Earthing will also help to prevent the possibility of fire due to fault. If a fault occurs between a phase conductor and earthed metalwork, there is a path for this fault current through the earth (Fig. 12.7a). This extra current will cause the operation of a fuse or circuit breaker to switch off the circuit and to isolate the fault.

Earth Fault Loop Test

The fault current mentioned above will only be large enough to operate the fuse or circuit breaker quickly if the impedance of the earth fault loop is low enough. Fault loop current is found by the formula.

$$I = \frac{V}{Z}$$

where V is supply voltage, and Z is earth fault loop impedance. Maximum values for the earth fault loop impedance for differing types of protective device in differing situations are given in the I.E.E. Wiring Regulations (Tables 41A1 and 41A2 of that publication).

The impedance of the earth fault loop cannot be easily measured using ordinary instruments, because a good deal of the loop is live whilst the supply is on. The supply cannot be switched off to facilitate measurement, because the supply conductors and the windings of the supply transformer are themselves part of the loop. Accordingly, a special instrument is used. An electronic device connects a low value resistor between phase and earth for a cycle or two of the supply voltage. The current flows for too little time to operate the protective device, but for long enough to be measured, together with the voltage, and to be used in a calculation based on $Z = V/I$. The result is displayed as the measured earth-fault loop impedance.

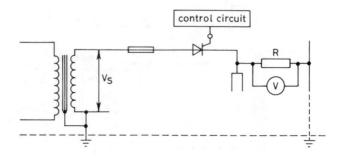

Fig 12.8 Simplified circuit for a phase–earth loop tester.

Earth Electrode Test

The earth electrode is the connection to the general mass of earth; it may vary from a complex grid of buried conducting plates or tapes to a simple rod driven into the ground. If the earth-fault loop impedance is too high, it is more than likely that the resistance between the electrode and the general mass of earth is too high.

A special instrument may be used for testing this resistance, or it can be tested using an a.c. supply, ammeter and voltmeter as shown in Fig. 12.9. Two temporary test electrodes are driven, with the first electrode Y at least twenty five metres from the electrode under test, X. The second test electrode Z, is driven exactly halfway between X and Y. X is disconnected from its earthing lead, and connected as shown in Fig. 12.9 so that an alternating current flows through the earth between X and Y, the value of this current being adjusted to a suitable value with R and read by the ammeter. The voltmeter reads the p.d. between X and Z, the latter being assumed to be outside the resistance area of both X and Y. The resistance area is the space around a live electrode in which the p.d. to the general mass of earth is not zero.

The earth electrode resistance is calculated from a simple Ohm's Law relationship, i.e.

$$\text{Earth electrode resistance} = \frac{\text{voltmeter reading}}{\text{ammeter reading}}$$

To ensure that the resistance areas of the electrodes X and Y have not overlapped, which would give a false result, electrode Z is moved, first six metres nearer to X (Z_1) and then six metres further from X (Z_2) than its first position. If the results of all three tests are the same, the areas do not overlap. If the test results are different this indicates that the areas do overlap, and the whole test must be repeated with Y (and hence Z) further away from X.

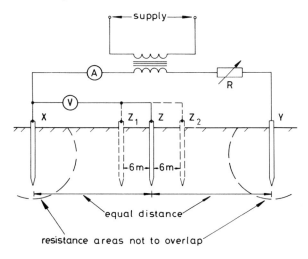

Fig. 12.9 Arrangements for testing on earth electrode

12.6 FUSES

In the event of excessive current in a circuit, its conductors and apparatus must be protected from overheating. A fuse is the simplest device which provides this protection.

The fuse consists of a thin wire placed in series with the circuit to be protected, so that it carries the circuit current. The wire is thick enough to carry normal load current without overheating, but, if the current exceeds the normal value, the fuse wire will melt, breaking the circuit. The thickness of the wire, or **fuse element**, is of obvious importance. If it is too thin, its resistance will be high, and the power dissipated in it by rated circuit current will raise its temperature to melting point. If it is too thick, its low resistance will dissipate little power, so that it may not melt even if the current becomes large enough to damage the circuit conductors or apparatus.

Similarly, the fuse enclosure is important. If the fuse element is open and unshrouded, it will quickly dissipate heat, and will require a large current to melt it. If enclosed so that heat cannot escape easily, it will melt at a lower current. There are two main types of fuse:

Semi-enclosed fuses

The fuse element consists of a wire of circular section which is connected to two screw terminals on the fuse carrier. The wire is usually made of an alloy containing 63% tin and 37% lead. The construction of a rewirable fuse carrier and base is shown in Fig. 12.10 (rewirable is the name commonly used for this type of fuse).

Semi-enclosed fuses have the advantages of low cost and simplicity, but have many disadvantages, including

(a) aging: oxidisation of the fuse elements often leads to failure of the fuse at rated current

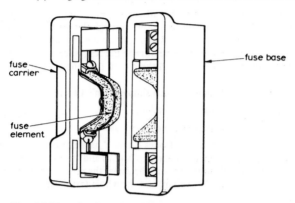

Fig. 12.10 Semi-enclosed fuse

(b) time delay in operation: this is significant when compared with h.b.c. (high-breaking capacity) fuses
(c) variations in fusing current due to the use of elements of different composition, and differing enclosures
(d) low rupturing capacity: in the event of a severe fault, the current may vapourise the element, and continue to flow in the form of an arc across the fuse terminals
(e) fire risk: the element becomes white hot as it operates
(f) the wrong fuse rating can be fitted, deliberately or by accident, by connecting an element of the wrong size: A 5 A circuit protected by a 60 A fuse in a 5 A carrier results in an obvious hazard.

High-breaking capacity fuses

This type of fuse was also known as a high rupturing capacity (h.r.c.) fuse. An h.b.c. fuse has its element enclosed in a cartridge of heatproof material, the cartridge being packed with chemically treated and graded quartz to prevent the formation of an arc. This type of fuse is nonaging, is very fast in operation, operates at a definite current for a given rating, and is capable of breaking very heavy currents. Owing to the enclosure of the element, there is no fire risk, and different current ratings are made with different cartridge sizes, often making it impossible to fit a fuse of the incorrect rating. The construction of a typical h.b.c. fuse is shown in Fig. 12.11.

H.B.C.. fuses are used increasingly, but their universal adoption is hampered by their high cost relative to rewirable fuses.

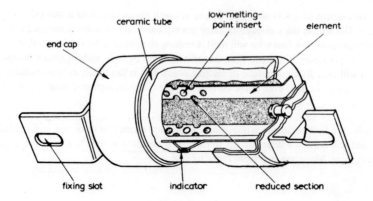

Fig. 12.11 High-breaking-capacity fuse

Practical supplies and protection 161

Fusing Factor

The current rating of a fuse is the current it will carry continuously without deteriorating. The minimum fusing current is the current which will cause the fuse to operate under given conditions in a given time. In BS88 the given operating time is four hours.

$$\text{Fusing factor} = \frac{\text{rated minimum fusing current}}{\text{current rating}}$$

It follows that the fusing factor must exceed one; the closer it is to one, however, the less likely is it that a fault current will fail to operate the fuse. Put simply, the lower the fusing factor, the better the fuse.

The time taken for a fuse to operate is a function of the current carried. The Wiring Regulations provide graphs of time/current characteristics for all types of fuse in Appendix 8.

Discrimination

A normal installation has a number of fuses connected in series. For example, the main fuse for a factory may be rated at 200A, with a 60A fuse feeding a power fuseboard which has 15A fuses protecting individual sub-circuits.

Obviously, a fault on the final sub-circuit should cause the 15A fuse to blow. If either the 60A or the 200A fuse blows, this will show a lack of discrimination on the part of the fuses.

Some fuses and circuit breakers are particularly fast in operation.

Care must be taken to ensure that such devices are not used to protect sub-circuits in which much slower-operating but lower-rated fuses are used, or the main fuse or circuit breaker may operate before that in the sub-circuit, putting a much wider area than necessary out of action. Such an eventuality would show bad discrimination.

12.7 CIRCUIT BREAKERS

The fuse element is a useful and simple protective device, but provides protection by destroying itself. After operation, it must be replaced, and demands the correct replacement h.b.c. fuse or fusewire, the use of tools, and the expenditure of time. A particular fuse can never be tested without its self destruction, and the results of the test will not necessarily apply to the replacement fuse.

The circuit breaker is an automatic switch, which opens in the event of carrying excess current. The switch can be closed again when the current returns to normal, because the device does not damage itself during normal operation. The contacts of a circuit breaker are closed against spring pressure, and held closed by a latch arrangement. A small movement of the latch will release the contacts, which will open quickly under spring pressure to break the circuit (Fig. 12.12).

The circuit breaker is so arranged that normal currents will not affect the latch, whereas excessive currents will move it to operate the breaker.

There are two basic methods by which overcurrents can operate, or 'trip', the latch:

(a) *Thermal tripping*: The load current is passed through a small heater, the temperature of which depends on the current it carries. This heater is arranged to warm a bimetal strip. This strip is made of two different metals, which are securely riveted or welded together along their length. The rate of expansion of the metals is different, so that, as the strip is warmed, it will bend and will trip the latch (Fig. 12.13). The bimetal strip and heater are so arranged that normal currents will not heat the strip to tripping point. If the current increases beyond the rated value, extra power is dissipated by the heater, $(P = I^2 R)$, and the bimetal strip is raised in temperature to trip the latch.

(b) *Magnetic tripping*: The principle used here is the force of attraction which can be set up by the magnetic field of a coil carrying the load current. At normal currents, the magnetic field is not strong enough to attract the latch, but overload currents operate the latch and trip the main contacts (Fig. 12.14).

162 *Practical supplies and protection*

There is always some time delay in the operation of a thermal trip, since the heat produced by load current must be transferred to the bimetal strip. Thermal tripping is thus best suited to small overloads of comparatively long duration. Magnetic trips are fast acting for heavy overloads, but uncertain in operation for light overloads. The two methods are thus combined to take advantage of the best characteristics of each; Fig. 12.15 shows a miniature circuit breaker having combined thermal and magnetic tripping.

Circuit breakers have many advantages over fuses, but are more expensive. They are made in a wide range of sizes, from the 5 A to 60 A miniature type for domestic use, up to industrial types capable of switching thousands of amperes.

The time taken for a circuit breaker to operate is a function of the current carried. The Wiring Regulations provide graphs of time/current characteristics for all three types of miniature circuit breaker (Appendix 8).

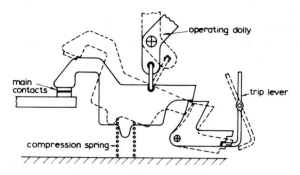

Fig. 12.12 Trip mechanism of miniature circuit breaker. Dotted positions are those of components after operation of trip lever. Even if operating dolly is held in 'on' position, trip lever will cause separation of contacts

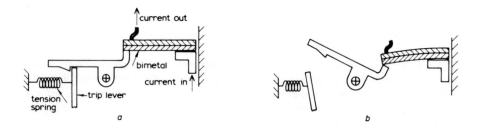

Fig. 12.13 Principle of thermal overload trip
  a In 'set' condition
  b In operated condition

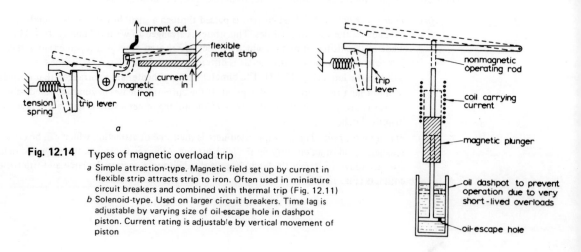

Fig. 12.14 Types of magnetic overload trip
  a Simple attraction-type. Magnetic field set up by current in flexible strip attracts strip to iron. Often used in miniature circuit breakers and combined with thermal trip (Fig. 12.11)
  b Solenoid-type. Used on larger circuit breakers. Time lag is adjustable by varying size of oil-escape hole in dashpot piston. Current rating is adjustable by vertical movement of piston

Practical supplies and protection 163

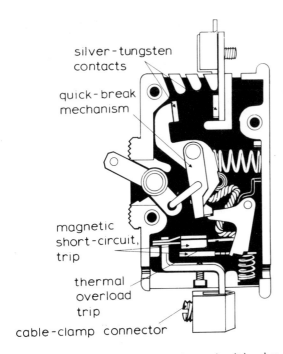

Fig. 12.15 General view of miniature circuit breaker

Earth Leakage Circuit Breakers

We saw in Section 12.5 that the earth fault loop impedance should be low enough to ensure that enough current flows in the event of a fault to blow the fuse or open the circuit breaker. In some installations, particularly where the supply is overhead and the sheath and armouring of an underground cable are not available for connection with earth, the measured impedance may be too high to ensure the operation of the fuse or circuit breaker, and an earth leakage circuit breaker must be used.

Residual Current Circuit Breaker

A simplified arrangement is shown in Fig. 12.16. The heart of the device is a magnetic core, which is wound with two main current-carrying coils, each with the same number of turns. These are connected so that the phase current in one coil provides equal and opposite ampere-turns (see section 6.4) to the neutral current in the other coil, as long as these two currents are equal.

Any earth fault current passes to the circuit through the phase, but does not return through the neutral The magnetic balance in the core is lost, and an alternating flux set up in it, the value depending on the amount of imbalance, and hence on the earth fault current.

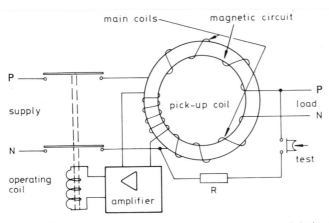

Fig. 12.16 Simplified arrangement for residual current earth leakage circuit breaker.

The flux induces an e.m.f. in the third winding, which is amplified and used to trip the breaker. A test button is required, and simulates a fault as shown.

The I.E.E. Regulations allow residual circuit breakers which operate when the earth current is as high as 2.0 A, but types operating with currents as low as 5 mA are available.

To prevent the possibility of the p.d. between earthed metal and earth becoming too high for safety, this type of protection can only be used when the operating current in amperes multiplied by the earth loop impedance in ohms does not exceed fifty. (V = IZ).

12.8 RISK OF FIRE AND SHOCK

Fire

Every day there are more than 200 fires in the UK which are judged to be caused by electrical faults. The majority of these faults are due to misuse of appliances, but almost a third are due to faulty installations.

There are three main causes for these fires:

(a) Overloads: more current-using appliances are connected to a circuit than it is designed to serve, and the current exceeds the circuit rating (Fig. 12.17). The circuit fuse will blow, or the circuit breaker operate, under these conditions, thus protecting the circuit. Danger arises only when the rating of the protection exceeds that of the circuit protected. For example, in Fig. 12.17 three 15 A loads have been connected to a 15 A circuit, blowing the fuse. A misguided handyman may replace the fuse, and, finding that it blows again, use a 60 A fuse. The overload current will not blow this fuse, but will certainly overload the circuit conductors, rapidly aging the insulation, and causing a fire from overheating or insulation failure.

(b) Short circuits: If a low-resistance fault occurs between phase and neutral, a very heavy current will result. This may cause arcing at the point of the fault, with resulting fire risk. Speedy operation of the protective device will prevent the buildup of heat which will cause a fire (Fig. 12.18).

(c) Earth fault: A low-resistance fault between phase and earthed metal will result in a fault current following the path indicated in Fig. 12.5. Failure of the protective device to open the circuit will again result in fire risk. Since the fuse or circuit breaker is current-operated, a sufficiently large current must flow to earth to operate it, and this will depend on a low value of resistance for the current path. It is possible, if the resistance of this path (the earth fault loop) is too high, for a current to flow which is high enough to cause overheating of the protective conductor without being high enough to blow the fuse or open the circuit breaker. Earth fault loops are tested to ensure that their resistance is low enough to enable the supply voltage to drive sufficient current through a low-resistance fault to operate the protection.

Shock

Section 12.12 will indicate in simple terms how an electric shock affects the human body. Over one hundred deaths occur in the UK each year as a result of electric shock. In addition, large numbers of persons are burned or otherwise injured as a result of shock. These shocks occur owing to two parts of the body coming into contact with conductors at differing potentials. They can be received in three ways:

(a) *shocks from phase to neutral*, owing to touching both conductors at the same time. Since the phase conductor is insulated and protected, such shocks are usually suffered by those foolish enough to tamper with installations, or by electricians working on live systems.

(b) *shocks from the phase conductor to the earth* (Fig. 12.6). Again, due to its insulation and protection, the phase conductor normally cannot be touched, but such accidents do happen. A not uncommon example is the dangerous practice of replacing a lamp, with the supply to the lamp-holder switched on.

The IEE Wiring Regulations refer to both these types of shock as due to 'direct contact'
Since the severity of the shock received is dependent on the voltage concerned, the shock danger can be reduced by using a lower voltage. Unfortunately, this results in increased current for a given power (P depends on *V and I*) with increased conductor-voltage drop and reduced efficiency of an appliance. One method commonly used is to employ a transformer with its secondary-winding centre tapped to earth (Fig. 12.19). The voltage to earth from either conductor is then only half that of the transformer secondary voltage.

(c) *shocks from noncurrent-carrying metalwork to earth.* In the event of a break in a protective conductor, the case of an appliance may be disconnected from earth. This fault is unlikely to be noticed until a second fault occurs from the phase conductor to the case, which will become live relative to earth (Fig. 12.7b).

This type of shock is said by the IEE Wiring Regulations to be due to 'indirect contact'.

Owing to the normal time delay of fuses and circuit breakers, no protection from shock is offered by them. The only protection is to ensure the efficient insulation of the circuit conductors, and the continuity of the earthing system.

The use of a residual current circuit breaker with a setting of 30 mA or less will often prevent death due to shock.

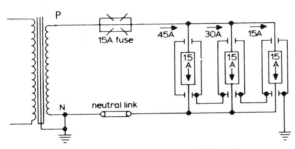

Fig. 12.17 Representation of overloaded circuit

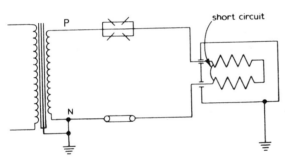

Fig. 12.18 Short circuit, giving very low circuit resistance, high current and causing operation of overload protection

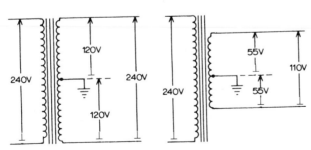

Fig. 12.19 Arrangement of transformers with earthed centre-tapped secondary windings

12.9 POLARITY

Section 12.8 has shown the danger of shock between a phase conductor and earth. If a circuit is to be switched off, or broken owing to operation of its protective device, it is ideal to open both conductors. Main switches are of the **double-pole** variety, breaking both phase and neutral conductors simultaneously.

For reasons of convenience in wiring and reduced cost, *single-pole* switches are often used in subcircuits.

It is essential that such switches be connected in the phase conductor, and not the neutral of a system with an earthed neutral. If properly connected, the part of the circuit controlled is made dead (and safe) when the switch is opened (Fig. 12.20a). Should the supply polarity be reversed, or a single-pole switch be connected in the neutral conductor, the switch will still switch off the part of the circuit it controls. This may seem to make the circuit safe, but as Fig. 12.20b shows, it is still alive, and any one touching a conductor (such as a fire element) will receive a shock if also in contact with earth.

For similar reasons, fuses and single-pole circuit breakers must be installed only in the phase conductors of systems with earthed neutrals. If the correct arrangement, called single-pole and neutral fusing, is used, the circuit is safe when the protection has operated (Fig. 12.21a). The neutral has no fuse or circuit breaker, but contains a solid link, which may sometimes be removed for testing. Nonearthed systems sometimes use double-pole fusing, with a fuse in both phase and neutral conductors. Such a system must *not* be used where the neutral is earthed. In the event of an overload or short circuit, it is quite likely that the neutral fuse will blow, leaving the live fuse intact. The circuit will not operate in this condition, but may still be dangerous (Fig. 12.21b).

12.10 SAFETY PRECAUTIONS

The electrical craftsman will necessarily be called on to work on electrical circuits. In the majority of cases, such circuits will be switched off, but the following precautions should always be taken:

(a) Never work on a 'live' circuit unless this is unavoidable. A few minutes of inconvenience is preferable to loss of a life.

(b) When working on a 'live' circuit, check insulated tools carefully to ensure sound insulation.

(c) Never rely solely on insulated tools for protection. Wear rubber-soled shoes and see that your free hand is not in contact with earthed metal. Often, it may pay to keep your free hand in your pocket.

(d) Test to ensure that the circuit *really is dead* before working on it. Do not rely on the circuit markings in the fuseboard.

(e) If the circuit control is not close to the place of work, make sure that the supply is not restored by someone else. Keep the fuses with you, and hang a warning notice at the control position.

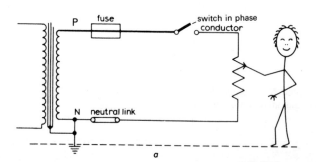

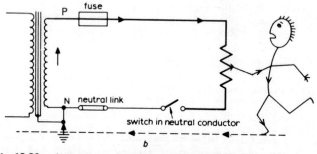

Fig. 12.20 Importance of polarity in single-pole switching
  a Switch in phase conductor. Appliance safe when switch is off.
  b Switch in neutral conductor. Appliance is still live although switched off

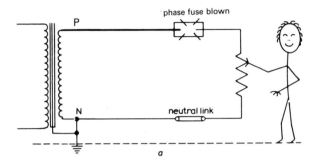

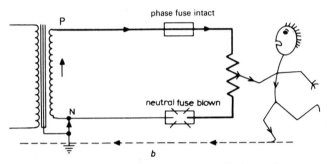

Fig. 12.21 Correct fusing for supply with earthed neutral

   a Single-pole and neutral fusing. Phase fuse 'blows' under fault conditions. No danger

   b Double-pole fusing. Neutral fuse may 'blow' under fault conditions, leaving circuit live

12.11 REGULATIONS

This chapter has indicated the major hazards associated with the use of electricity, and outlined the more important methods of ensuring safety. In practice, there are many other hazards, and other regulations apply to electrical installation of various types. The major regulations are

I.E.E. Wiring Regulations

The IEE Wiring Regulations, published by the Institution of Electrical Engineers, are the most important and far reaching of the rules applying to electrical installations. These regulations are kept up to date by issuing new editions and amendments from time to time.

The IEE Wiring Regulations are of the greatest importance, and everyone connected with electrical installations should possess a copy and become familiar with it. Despite the length and complexity of the IEE Wiring Regulations, they cannot cover every possible contingency, and the advice and expertise of a professionally qualified electrical engineer will sometimes be required. Further guidance on good practice will be found in the relevant British Standard Codes of Practice.

Electricity Supply Regulations

Copies of these Regulations, issued by the UK Government, can be obtained from Her Majesty's Stationery Office (HMSO). They apply to the distribution and supply of electrical energy. Compliance with the IEE Wiring Regulations is required before an installation is connected to its supply.

Electricity (Factories Act) Special Regulations

These regulations, also obtainable from HMSO, are issued by the Department of Employment & Productivity, and compliance with them is a legal requirement for factory premises. The Health & Safety Inspectorate has the duty to ensure that factory electrical installations comply with them in all respects.

Other special regulations

Other special regulations apply to particularly hazardous situations such as mines, quarries, and places of public entertainment. Details of full titles and sources are given in the IEE Wiring Regulations.

British Standards

Quite clearly, if an installation is to be safe for the whole of its lifetime, it must not only be installed correctly by the electrician, but also contain only materials which are safely constructed. For instance, a switch with a live metal dolly obviously would be dangerous, and no one would make one for use at mains voltages. A switch with a live metal dolly insulated by some form of plastic would appear to be safe, but would be just as dangerous as the switch with the bare metal dolly, if the plastic insulator broke. The British Standard for such a switch would ensure that it was constructed so that the insulation would not be broken in normal service. It is more likely, however, that it would specify that any metal reinforcement used for the dolly should not become live. British Standard Specifications have been produced for nearly all types of electrical equipment, and are listed in the IEE Wiring Regulations.

Health & Safety at Work etc. Act, 1974

This Act is written in general terms, and applies to many other hazards as well as electrical ones. It places a statutory duty on craftsmen to take care for their own safety as well as that of other people, and may lead to heavy fines or even imprisonment if infringed.

12.12 ELECTRIC SHOCK

There can be few among us who have not at some time received an electric shock. The vast majority of shocks are so slight as to cause only minor discomfort, and it is this fact that has contributed to the general lack of appreciation of the real dangers involved.

An electric shock is the passage of an electric current through the body. The amount of current which is lethal varies from person to person, and also depends on the parts of the body in which it flows. To understand why we are 'shocked', we must realise that every movement we make, conscious or unconscious, is produced by muscles reacting to minute electric currents generated in the brain. These currents are distributed to the correct muscles by the 'conductors' of the nervous system.

If a current much larger than the one usually carried is forced through the nervous system, the muscles react much more violently than normal, and hence we experience the 'kick' associated with an electric shock. If, in addition, the nerves carry the excess current to the brain, it may destroy, or cause temporary paralysis of, the cells which generate the normal currents. Destruction of these cells means almost instant death, as the heart muscles cease to operate, and no blood is circulated. Paralysis results in unconsciousness, but, if the lung muscles are not operating, death from suffocation will follow in a few minutes.

Severity of shock

The severity of shock depends on the amount of current flowing in the body. As seen in Chapter 1, the amount of current flowing increases as the voltage applied increases, and decreases as circuit resistance increases. Thus, in identical circumstances, a worse shock will be received from a high voltage than from a low voltage. Many circuits, such as those for bells and telephones, are operated at low voltage so that the risk of shock is removed. High-voltage shocks are often accompanied by severe burns.

For a given voltage, the severity of the shock received will depend on circuit resistance, which is made up of the following parts:

(a) *Resistance of the installation conductors*: These form such a small proportion of the total resistance that usually they may be ignored.
(b) *Resistance of the body*: This varies considerably from person to person and, for a particular person, with time. As the body itself is made mainly of water, its resistance is quite low, but it is covered by layers of skin which have high resistance. It is in the resistance of the skin that the main variations occur. Some people have a naturally hard and horny skin which has high resistance, and others have soft, moist skin of low resistance. If the skin is wet, the moisture penetrates the pores, giving paths of low resistance.
(c) *Contact with the general mass of earth*: The body is normally separated from the conducting mass of earth by one of more layers of insulating material, i.e. shoes, floor coverings, floors etc. It is the resistance of these insulators which normally prevents a shock from being serious. For instance, a man wearing rubber-soled shoes standing on a thick carpet over a dry wood floor can touch a live conductor and feel nothing more than a slight tingle. The same man standing on a wet concrete floor in his bare feet would probably not live to describe his sensations on touching it!

This account is over-simplified, but will serve to give an indication of what occurs. Very little information is available in terms of actual figures, because, clearly, this is a subject which does not lend itself to practical experiment.

12.13 ARTIFICIAL RESPIRATION

As explained in the Section 12.12, severe electric shock is often accompanied by a form of paralysis of the nervous system. The heart continues to beat, and the victim is not yet dead, but as breathing has stopped, he will soon die if no action is taken.

Artificial respiration keeps the patient's lungs working until his own system can recover and take over. It is very simple to apply, and is basically the same as that used in drowning cases. Every person should be proficient in its use.

Precautions

If you witness an accident due to electric shock resulting in unconsciousness, or find an unconscious colleague, remember the following points:

(a) Ensure that the patient is not still in contact with the electrical system. If he is, and you touch him, you may receive a shock, too. Switch off the supply, or, if this is not possible, drag him clear with dry clothing or some other insulator.
(b) Do not waste time trying to find out if he is still alive. It is vital to start respiration *at once*.
(c) Summon assistance, namely doctor, ambulance etc., at once, but do not delay artificial respiration to do so. If you are on your own, shout for help periodically.

Types of artificial respiration

For a number of years the recommended system has been that known as the **Holger-Nielsen** type. This involves lying the patient face down on the floor, alternately applying pressure to the shoulder blades and pulling forwards on the elbows to force air in and out of the lungs. This system has saved many lives, but is relatively slow, and complications can be caused if the patient has broken bones or internal injuries. Such injuries are often associated with unconsciousness due to shock as a result of falling

Another type of artificial respiration which has grown in popularity is the **rocking-stretcher** method. The patient is put on a pivoted stretcher, which is rocked slowly up and down. Since the stomach and intestines can move within the body, they press air out of the lungs when the head is down, and draw air into the lungs as the head is lifted. If no special stretcher is available, this method can be performed by two operators. The need for special equipment or more than one operator is an obvious disadvantage, although injuries are seldom worsened.

The **mouth-to-mouth** system, sometimes called the **kiss of life**, appears to be accepted generally as the quickest and most effective system.

Mouth-to-mouth artificial respiration

The patient should be quickly laid on his back, *with the head tilted as far backwards as possible*. This will open and straighten the air passages. Clear the mouth of foreign objects such as false teeth, vomit, etc., and make sure that the tongue is not blocking the airways. Close the patient's nose by pinching his nostrils, place your mouth firmly over his, and blow. When his chest is inflated, remove your mouth, drawing in breath as you do so. The patient's chest will deflate, after which the cycle should be repeated. This process should take place 10–12 times each minute, that is, once every five or six seconds. Counting the seconds to yourself may help you to prevent slowing down, which can prove fatal.

An alternative method is **mouth-to-nose** artificial respiration, which is easier in some cases. The victim's mouth should be closed with the thumb, and your mouth placed firmly over the patient's nose for the blowing operation. If the chest is not inflated, check the head position and try again. Failure on the second attempt means that you must change to the mouth-to-mouth system at once.

12.14 SUMMARY OF FORMULAS FOR CHAPTER 12

For a delta or mesh connection,

$$V_L = V_P \qquad I_L = \sqrt{3} I_P \qquad I_P = \frac{I_L}{\sqrt{3}}$$

For a star connection,

$$I_L = I_P \qquad V_L = \sqrt{3} V_P \qquad V_P = \frac{V_L}{\sqrt{3}}$$

170 Practical supplies and protection

where V_L = line voltage (voltage between supply lines)
V_P = phase voltage (voltage across one load)
I_L = line current (current in a supply line)
I_P = phase current (current in one load)

12.15 EXERCISES

1. Show by means of simple diagrams what is meant by
   (a) a single-phase supply,
   (b) a 3-phase supply.

2. What is meant by the phase conductor of a 2-wire a.c. supply? (NCTEC)

3. Draw a sketch of the earth fault loop, naming the parts.

4. What are the advantages and disadvantages of connecting the neutral conductor of a supply system to earth?

5. A portable drill is connected to the supply by a 2-core-and-earth cable, and is controlled by a single-pole switch. The mains terminals are *phase*, *neutral* and *earth*. In the supply terminal box are a link, a fuse and an earth terminal. Make a labelled diagram of connections. Explain concisely
   (a) why you have included the switch in a particular lead
   (b) why a link and not a fuse is used in one lead
   (c) how the earth-lead inclusion is intended to protect the user. (C & G)

6. Why is a fuse regarded as a protective device in an electric circuit? (NCTEC)

7. What are the fusing arrangements required in a single-phase a.c. supply with
   (a) earthed neutral,
   (b) no earth?
   Explain the reasons for this difference. (NCTEC)

8. Describe the construction of a semi-enclosed fuse. List the advantages of this form of protection compared with an h.b.c. fuse.

9. Show by means of a clearly labelled drawing the construction of an h.b.c. fuse. What are the advantages of this fuse?

10. Using simple diagrams, explain the principles of operation of both thermal and magnetic overload trips. For what type of overload is each particularly suited?

11. Make a neat, labelled sketch to show the construction and operation of an adjustable solenoid-and-plunger-type overcurrent trip device. (C & G)

12. What are the two major risks associated with the use of electrical energy, and how do they occur?

13. What steps are taken in the design of an electrical installation to ensure safety from fire?

14. What steps are taken in the design of an electrical installation to ensure safety from shock?

15. Describe by means of circuit diagrams and brief explanations what precautions against electric shock to users are necessary for
    (a) portable electric power tools in a factory
    (b) lighting and heating in a domestic bathroom
    (c) electric-bell circuits supplied from the mains. (C & G)

16. With the aid of a simple sketch, explain the meaning of the term 'short circuit'. (NCTEC)

17. (a) Portable electrical appliances in a factory are often fed from a 240 V/110 V transformer with a centre tap on the secondary winding. Draw a diagram to show how this is connected to the appliance, and explain how it reduces the risk of electric shock. (C & G)

18. Draw a diagram of a 415 V, 3-phase, 4-wire distribution system. Indicate on the diagram the connections for supplying the following loads:
    (a) a single-phase lighting load
    (b) a 3-phase motor
    (c) a 415 V, single-phase electric welder. (ULCI)

19. What are the standard values of the voltages available from a 3-phase, 4-wire, a.c. supply? Give the standard colour code for the four wires. (NCTEC)

20 Explain briefly how the risk of electric shock is reduced by the use of the following:
   (a) a fuse in the phase line and a solid link in the neutral line
   (b) a transformer to supply a bell and bell-push circuit from domestic supply. (C & G)

21 List the safety precautions to be taken when working on electrical circuits. What additional precautions become necessary if isolation of the circuits from the supply is not carried out?

22 Explain the action you would take if you found a colleague who was unconscious as a result of a shock. Assume that the colleague may still have been in contact with the live supply.

Chapter 13

Cables and enclosures

13.1 INTRODUCTION

So far in this book we have considered some of the basic theories applying to the uses of electricity, and have mentioned the dangers involved in its use. Before electricity can be utilised, it has to be transmitted from the generating station to the points of use. Cables are used for this purpose. Overhead, or armoured underground, cables will be used for mains supplies, and insulated cables with one or other of many types of mechanical protection will be used for the electrical installation of a building. This chapter will examine the types of cable and methods of protection in use, showing the applications.

13.2 CONDUCTOR MATERIALS AND CONSTRUCTION

Basic electrical conductors have already been considered in Section 1.4, showing that, from a purely electrical point of view, silver is the best conductor, but that its poor mechanical properties and high cost rule it out as a cable conductor material.

Copper

Copper is second only to silver as an electrical conductor, is easily drawn into wires, and is comparatively strong, but may be softened (annealed) to make it easier to bend. All small conductors, as well as many for heavy power cables, are made of copper. However, the price of copper is high, and changes from day to day.

Aluminium

Although aluminium is more than half as resistive again as copper, its density is less than one third that of copper, so an aluminium conductor, although larger, will have half the weight of an equivalent copper conductor. The disadvantages of lower strength, higher thermal expansion, and rapid oxidisation which necessitates special jointing techniques, are offset by the low and stable price. Aluminium conductors are almost always used for heavy-current overhead lines (often with a steel core for extra strength), and very widely for power cables. At the time of writing, aluminium cables of cross-sectional area less than 16 mm^2 are not allowed, although it is possible that this Regulation will be changed in due course. Table 7 shows the relationship between copper conductors and equivalent aluminium conductors.

Table 7 Standard sizes of copper cables with comparable aluminium sizes

Copper-conductor c.s.a.	Aluminium conductor (of equivalent current rating) c.s.a.
mm^2	mm^2
1·0	1·5*
1·5	2·5*
2·5	4 *
4	6 *
6	10 *
10	16
16	25
25	35
35	50

* The use of these cables does not comply with IEE Regulations

Solid and stranded conductors

Some small cables, mineral-insulated and aluminium power cables have single strand or solid conductors. The majority of cables, however, have stranded conductors to make them more flexible for installation purposes. The standard arrangements are for one, seven, 19 or 37 strands (Fig. 13.1). Flexible cords and flexible cables have more numerous but smaller strands (Table 8).

Table 8 Stranding of copper cables

Cross-sectional area, mm²	The first number indicates the number of strands; the second number gives the diameter of each strand in mm	
	Cables	Flexibles
0·5	—	16/·20
0·75	—	24/·20
1·0	1/1·13	32/·20
1·5	1/1·38	30/·25
2·5	1/1·78	50/·25
4	7/0·85	56/·30
6	7/1·04	84/·30
10	7/1·35	80/·40
16	7/1·70	126/·40
25	7/2·14	196/·40
35	19/1·53	276/·40

It is possible to obtain standard wiring cables up to 2·5 mm² with multiple, as well as single, strands of conductor. Such cables are used in situations where their extra flexibility is useful.

When copper and aluminium conductors are drawn out into strands they become **hard-drawn**, in which condition they are stiff and hard. They can be **annealed** by heat treatment, when they become comparatively soft and pliable. Continuous bending of the conductors, particularly if the strands are large, will result in **work-hardening** and possible cracking.

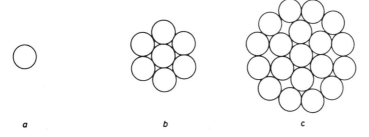

Fig. 13.1 Conductor sections
 a Single strand
 b Seven strands
 c 19 strands

Heating cables

Most electrical conductors are constructed so that they have low resistance. This results in small power losses and low temperature rises within the cable itself. Sometimes however, cables are used for heating, and the conductors must then have a higher resistance. Such conductors are usually alloys containing some combination of copper, kumunal, iron, steel, nickel and chromium. Heating cables are used for underfloor heating, keeping roads and ramps clear of snow, keeping gutters free of ice, soil heating for horticultural purposes etc.

13.3 CABLE INSULATORS

A very wide range of cable-insulation material is now in use, some of the more important being listed with their properties below:

Polyvinyl Chloride (p.v.c.)

A basic plastic material which can be made in many forms. It is robust, chemically inert, has good aging and fire-resisting properties. It is more resistant to weathering and sunlight than rubber, but will harden and crack in the presence of oil or grease. High temperatures lead to softening and possible insulation failure, whereas at low temperatures, p.v.c. may become brittle. Very little absorbtion of water will occur, although p.v.c. is inferior to p.c.p. in this respect. p.v.c. is widely in use as a sheathing material as well as for insulation. Can only usually be used where conductor temperature does not exceed 70°C.

Vulcanised Rubber (v.r.)

A good electrical insulator with high flexibility. Subject to rapid aging and cracking due to weathering and exposure to sunlight. Softens and becomes sticky in the presence of oils and greases. Ages rapidly at high or low temperatures. Not widely used but available in a form which will operate safely up to 85°C.

Polychloroprene (p.c.p.)

Polychloroprene, also known as Neoprene, is a plastic material of lower strength and lower insulation resistance than p.v.c. However, it is more resistant than p.v.c. to weathering, and to attack by oils, acids, solvents, alkalis and water. p.c.p. is more elastic than p.v.c., and this elasticity is not affected by increased temperatures. Although more expensive than p.v.c., p.c.p. is used as a cable insulation and sheathing material for conditions where p.v.c. would not be suitable.

Silicone rubber

Silicone rubber is a synthetic material with many of the advantages of natural rubber. It has good weathering properties, and will resist attack from water and mineral oils, but not from petrol. It remains elastic over extremes of temperature ($-70°$C to $150°$C).

Butyl rubber

Butyl rubber is another synthetic material which is less expensive than silicone rubber but has similar advantages. It remains flexible over the range $-40°$C to $85°$C, but will burn readily once ignited.

Ethylene propylene rubber (e.p.r.)

E.P. rubber is generally similar to butyl rubber, but with improved properties. It is resistant to heat, water, oil and sunlight, and is suitable for direct burial, exposure to weather and contaminated atmospheres. It is used to insulate power cables which have an increased current rating owing to its heat resistance.

Chlorosulphonated polyethylene (c.s.p.)

This material is mainly used for sheathing cables insulated with other plastics. It is capable of operating over the temperature range $-30°$C to $85°$C, and has excellent oil resisting and flame-retarding properties. It is very tough, and very resistant to heat and water. This is the sheathing material which is called 'h.o.f.r.' (heat resisting, oil resisting and flame retarding) in the IEE Regulations.

Magnesium oxide

This material is in the form of a white powder, and is invariably used in mineral-insulated cables. It also finds application in some heating elements. Magnesium oxide is nonaging, and will not burn, but it is very hygroscopic (absorbs moisture from the air), when it loses its insulating properties. It is unaffected by high temperatures, and is a good conductor of heat.

Cross-linked polyethylene (XLPE)

This is a polmeric insulation (like e.p.r. above) and is used particularly for high voltage power cables. In this application it shows cost advantages over oil-filled and pressurised cable insulation, not least in the simpler jointing techniques which may be used.

Paper

Dry or oil-impregnated paper, wound in long strips over the conductors, is the most common method of insulating underground cables. As long as the paper is kept dry (usually by sheating with lead alloy) its insulating properties are excellent.

Glass fibre

Glass fibre, impregnated with high temperature varnish, is used to insulate some flexible cords used for high-temperature lighting applications.

13.4 BARE CONDUCTORS

Bare conductors are not covered with insulation. Those intended for use at voltages which are above a safe level will be normally supported on insulators. The IEE Regulations approve the application of bare conductors where they are used as

   (a)   the external conductors of earthed concentric systems. This type of wiring uses an inner, insulated core for its phase conductor, and the outer conductor or sheath as a combined neutral and earth conductor

(b) the conductors of system working at a safe voltage. An example is where galvanised-steel wire is buried in the soil and fed from a transformer at low voltage to provide soil heating

(c) protected rising-main and busbar systems. These systems use bare busbars enclosed within a steel trunking and supported on porcelain or plastic insulators. Provision is made for supplies to be tapped off at intervals. The system is often used as a mains-supply distributor for multi-storey flats or for factories. See Fig. 13.2.

(d) as current-collector wires for overhead cranes and the like. The conductors must then be well out of reach, and clearly labelled to indicate the danger.

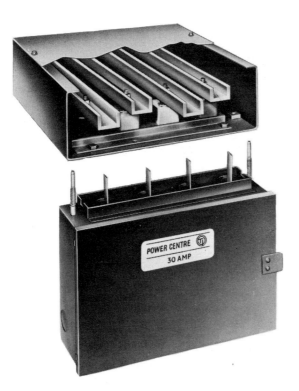

Fig. 13.2 Busbar trunking with plug-in type fused tapping box

13.5 PLASTIC AND RUBBER-INSULATED CONDUCTORS

These cables are insulated, and, in some cases, may also have an overall braiding to protect them during installation. IEE Regulations demand that they be protected against mechanical damage, and so they must be completely enclosed within a duct, trunking or conduit. The minimum bending radius for insulated conductors is three times the overall diameter for conductors not exceeding 10 mm^2. Fig. 13.3 shows typical constructions of insulated conductors.

Fig. 13.3 Insulated conductors: a p.v.c.-insulated, b Elastomer insulated, braided and compounded

13.6 SHEATHED WIRING CABLES

For most domestic and commercial installations, the least expensive system includes the use of sheathed cables. These consist of conductors insulated with p.v.c. laid up singly or with others, and covered by a protective sheath. The sheath may consist of p.v.c. or p.c.p.; p.c.v.-sheathed cables are the most widely used, p.c.p. finding particular application for outdoor installations (Fig. 13.4).

Sheathed cables are fixed by clips pinned to the surface, or may be buried in plaster walls, with or without the added protection of a conduit or capping. Where run under floorboards, the cable must pass through holes which are at least 50 mm below the top of the joist to prevent damage by nails.

Fig. 13.4 P.V.C.-insulated p.v.c.-sheathed cables
    a Single-core, stranded
    b Single-core, single-wire, with earth
    c 3-core, with earth, single-wire

Joint boxes are used to connect sheathed cables, and, as with other terminations, the sheath must be continued right into the box, so that no unprotected conductors are exposed to damage.

A special sheathed cable with a further protective braiding and compounding is used on farm installations, where the corrosion problem is acute. Special joint boxes and accessories are available for this system.

Sheathed cables should not be manipulated or installed when the temperature is below 0°C, or there is a danger that the sheath or insulation may crack. If cables are tightly bent, the sheath may facture; the minimum bending radius should never be less than three times the overall cable dimension and is larger for cables with conductors larger than 10 mm².

13.7 MINERAL-INSULATED CABLES

This cable consists of solid cores of high-conductivity copper embedded in highly compressed magnesium oxide, the whole being contained in a solid-drawn copper sheath. This sheath may be further protected by an overall covering of p.v.c., which will prevent deterioration in the presence of moisture and of most chemicals (Fig. 13.5).

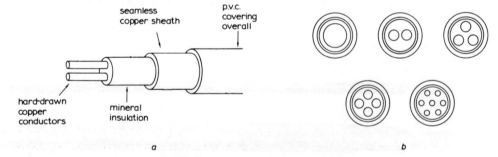

Fig. 13.5 Mineral-insulated cable
    a Components of 2-core m.i. cable with p.v.c. sheath overall
    b Standard conductor arrangements

Standard cables are available with one, two, three, four or seven cores, with voltage ratings of 240 V, 440 V and 660 V. Mineral-insulated cables are nonaging, have high mechanical strength, are small in size, are very resistant to corrosion, have excellent electrical properties, are waterproof, can work at high temperatures, and are suitable for use in flameproof installations. They must be terminated with special seals (Section 13.9) to prevent the insulator becoming moist, when it will lose its insulating properties.

The hard-drawn sheath may crack owing to work-hardening if bent too often, or if bent too tightly. Minimum bending radius must not be less than six times the cable diameter.

There is often no identification of the cores of a multicore m.i. cable, which must be checked through with a continuity tester and marked with coloured sleeves.

The copper sheath of this cable must be kept separate from the metalwork of water and gas systems, or, where this is not possible, must be securely bonded to them.

Mineral-insulated cables with aluminium conductors and sheath are available.

13.8 ARMOURED CABLES

An armoured cable is one where the conductors and insulation are protected by a layer or layers of steel wires or tapes. This armouring protects the cable from mechanical damage while it is in service, as well as giving added strength to withstand handling during installation.

The conductors of these cables are made of copper or aluminium, the former always stranded, but the latter often solid. Conductors are often sector-shaped (Fig. 13.6) to fit together more closely.

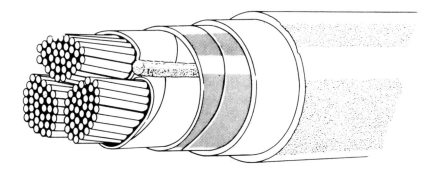

Fig. 13.6 3300 V paper-insulated cable

Dry or impregnated paper has been in use for many years as the most usual insulator for these cables, with a lead sheath to keep out moisture. In recent years, p.v.c. has been used for both insulation and sheath of lower voltage cables, with XLPE as insulation for high-voltage types. A soft bedding of jute or other suitable material is sometimes applied over the sheath to serve as a base for the steel armour.

Steel wires are given a twist, or 'lay', along the cable so that they fit neatly into the bedding. Steel-tape armouring may be wound round the cable in a spiral, or may be in the form of longitudinal strips. The armouring is usually, but not always, protected from corrosion by an overall 'serving' of impregnated jute, or of p.v.c.

Power cables of these types find wide application for underground electrical supplies, as well as for the heavy submain distributors in some large buildings. They are also used for circuit wiring in factories where the likelihood of mechanical damage is great.

Paper-insulated cables are slightly more expensive, and are more difficult to joint (Section 13.9) than the p.v.c.-insulated types, but are able to carry more current for a given size, because increased conductor temperature does not affect the paper insulation. Figs. 13.6, 13.7 and 13.8 show some examples of armoured cables.

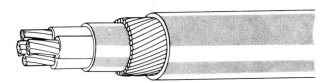

Fig. 13.7 P.V.C.-insulated and -sheathed steel-wire-armoured cable

13.9 CABLE JOINTS AND TERMINATIONS

Cables must be terminated by connecting them at control gear, accessories etc., but joints along the run of a cable should be avoided wherever possible. Terminations and joints must be electrically and mechanically sound, the principles being that electrical resistance should be low, and insulation resistance high.

Cables of differing types require different treatments, terminations and joints in some of the most common cables being detailed below.

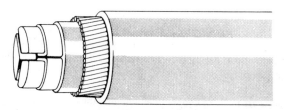

Fig. 13.8 P.V.C.-insulated and -sheathed steel-wire-armoured cable with solid-aluminium conductors

Rubber and plastic-insulated cables

There is no need with these cables to seal the cable ends against the ingress of moisture, and the pinching screw is the most common type of terminal (Fig. 13.9). Some types of single-strand conductor, particularly of aluminium, have been found to work loose in time owing to deformation of the conductor material. To prevent this difficulty, the pinch screw should not be overtightened; the pressure-plate type of terminal gives good results with single or stranded cores.

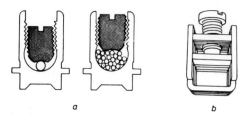

Fig. 13.9 a Well designed pinch-screw-type terminal
b Pressure-plate-type terminal

Where larger cables must be terminated, for example in switchgear, the grip type of cable clamp may be employed, gripping the whole of the conductor and not causing damage.

It is still fairly common for large cables to be connected to switchgear or busbars by fixing lugs to the conductor ends. The traditional method is to sweat the lug solidly to the conductor using tin-man's solder and a noncorrosive flux (Fig. 13.10). An alternative method is that which fixes the lug to the cable by means of pressure and indentation. The conductor is inserted into the lug, and an indentation made by

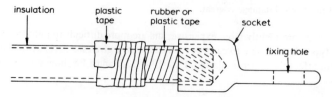

Fig. 13.10 Socket soldered to cable. Insulation removed for soldering or damaged by heat is made good with tape

using a hand tool or a hydraulic press. The pressure compacts the lug and the cable into a single solid mass, so that the joint is stronger than the cable itself (Fig. 13.11). This type of termination is quicker and cheaper than the sweating method, and causes no damage to the insulation. The compression lug is used with both copper and aluminium cables, whether stranded or solid. It is particularly useful with aluminium cables, which are difficult to solder, owing to the very rapid oxidisation of surfaces after cleaning.

Aluminium may corrode when in contact with other metals, especially in the presence of moisture, or may cause corrosion of the other metal. Aluminium-to-aluminium joints should always be made when possible, but, where this cannot be done, special precautions may need to be taken. These include electroplating of the other metal, or coating it with a layer of grease or compound before connection. Corrosion is particularly likely when aluminium is in contact with a metal having a high copper content such as brass.

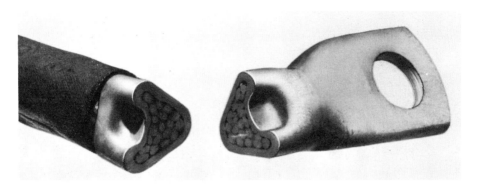

Fig. 13.11   Pressure lug cut through to show effect of 'cold welding'

Mineral-insulated cables

The mineral insulation of these cables is very hygroscopic. This means that the insulation will absorb moisture from the air. Wet insulation of this type loses its insulating properties, so the cable ends must be sealed to prevent moisture getting in.

The sealing pots used are screwed or wedged on to the cable sheath, and enclose the moisture-resisting compound, which is compressed by crimping a disc into position in the pot end. The disc also secures the p.v.c. sleeves used to cover the bared copper conductors. The completed seal is either enclosed in a gland, an 'exploded' view of one type being shown in Fig. 13.12, or can be accommodated in a 'glandless box' (Fig. 13.13). The conductors are terminated in pinch- or clamp-type terminals, or by means of cone grip, sweated or pressure lugs.

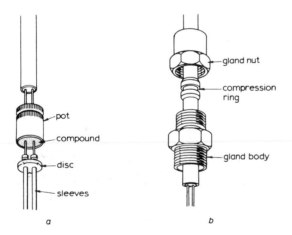

Fig. 13.12   Terminating mineral-insulated cable
  a Seal
  b Gland

Flexible cords and cables

Flexibles are usually connected to portable appartus and plugs, the normal method being by wrapping the conductor round a captive screw under a washer. If the conductor is wrapped in a clockwise direction, it will tighten, and not push out as the screw is tightened. Pinch-type terminals are used in some cases, but extra care is needed in the manufacture, as any sharpness on the end of the pinching screw may sever the small conductor strands.

Lugs of both sweated and pressure types can be used with flexibles. With sweated lugs, special care is needed; if the heat is applied for too long, solder will creep up the conductor between the strands. If it does so, that section of the conductor will become rigid, and a break will occur in due course owing to bending at the junction of the flexible and rigid sections of the conductor.

Fig. 13.13 Glandless boxes for use with m.i. cables

Armoured cables

Armoured cables are terminated or joined with the aid of special glands or sealing boxes. P.V.C.-insulated and -sheathed types use glands, of the type shown in Fig. 13.14. The sealing ring shown over the sheath at the end of the gland may be omitted if the gland is for use indoors, because the p.v.c. insulation is not hygroscopic.

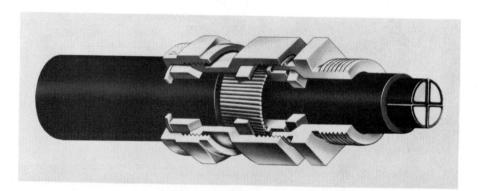

Fig. 13.14 Gland for use with p.v.c.-sheathed armoured cable

Paper-insulated cables have a hygroscopic insulation which must be sealed. Joints are made by the compression method or by sweating the conductors in copper ferrules, the insulation being made good with tape. The joint or termination is then enclosed in a box, which is filled with a moisture-proof insulating compound. There are very many arrangements for joints and terminations; a typical terminal box, with paper-insulated conductors joined to p.v.c. tails, is shown in Fig. 13.15.

13.10 CONDUITS

If, as in most electrical installations, it is necessary for a number of conductors to follow the same route, it is often convenient and economical to use insulated-only conductors and to enclose them in a common protective tube. Such a tube is a conduit, and is most usually made of steel, although plastic, aluminium and copper conduits are also available.

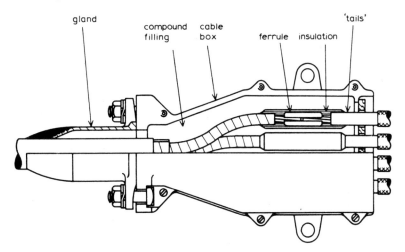

Fig. 13.15 Sealing box for paper-insulated cable

The conduit system is erected complete before the cables are pushed or drawn into it, inspection boxes often being used if long runs or numerous bends are involved.

Cable capacities of conduit and trunking

The number of cables of a given size which can be fed into a certain conduit obviously will be limited by physical considerations, but will be further reduced by electrical requirements. For example, if cables are tightly packed, insulation will be compressed, its thickness reduced, and its effectiveness impaired. More importantly, cables become warm in service owing to the current they carry. If this heat cannot escape, and too many cables in a conduit make heat transfer more difficult, the cables will overheat, and insulation may be damaged.

To prevent overcrowding of conduits, IEE Regulations provide tables for common cable, conduit and trunking sizes. For other cases they specify that the cables must not take up more than 45% of the space available within the conduit. This is specified by giving a figure for the 'space factor':

$$\text{Space factor} = \frac{\text{cross-sectional area of cables}}{\text{internal cross-sectional area of conduit}} \times 100\%$$

and must not exceed 45%.

It is important to appreciate the fact that the cross-sectional area of cables includes the insulation, so that the overall cross-section of a $2 \cdot 5 \text{ mm}^2$, 600 V, single-core cable is actually $9 \cdot 6 \text{ mm}^2$

Light-gauge-steel conduit

These conduits are often used in straight lengths with rubber-bushed ends for the protection of sheathed cables where these are buried in plaster. They are made of thin section material, often no thicker than 1 mm, which is bent into a circular or oval shape with the edges butted, and often not joined together as in heavy-gauge conduit. Circular light-gauge conduit may be erected into a complete system by the use of grip-type boxes, but, since it cannot be bent successfully, its use is very limited.

Heavy-gauge-steel conduit

As the name implies, this conduit is thicker-walled than the light-gauge types, with wall-thicknesses of $1 \cdot 65$ mm and above. The conduit is erected by threading it, some conduit connections being shown in Fig. 13·16. Joints are accomplished by means of couplers (sockets) (Fig. 13.16a), with circular or rectangular boxes providing for drawing-in of cables or mounting accessories. Connections to boxes may be by means of threaded spouts provided, or using a male bush and socket or a female bush to connect to a clearance hole (Fig. 13.17).

In most industrial applications, there is no need to hide conduit, and it is fixed to the surface of walls and ceilings by means of saddles. To allow for easy drawing-in of cables and to prevent damage to the insulation, conduit ends must be butted up in couplers, and all burrs must be removed.

Conduits are often chosen to accommodate cables for installations in concrete buildings, such as flats and office blocks, and here the installation must be hidden. This is frequently accomplished by burying the

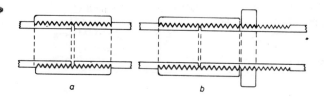

Fig. 13.16   Conduit connections.
  a Conduits tightly butted in socket
  b Running socket with locknut

conduit in the concrete itself, special outlet boxes being available for this purpose (Fig. 13.18). Heavy-gauge conduit is bent where necessary using a bending machine; the bending radius must never be less than 2·5 times the outside diameter of the conduit. Where a long vertical run of conduit is involved, cables must be supported within the conduit at intervals of not more than 3 m to prevent strain. Fire barriers must be placed within a conduit where it may otherwise allow the spread of fire. Heavy-gauge-steel conduit usually forms the protective conductor.

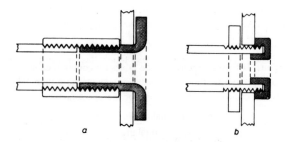

Fig. 13.17   Conduits connected to clearance hole
  a Male-bush and socket method
  b Female-bush and locknut method

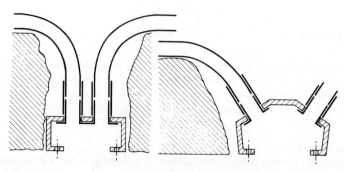

Fig. 13.18   Standard and angle-pattern looping-in boxes

Plastic conduit

Plastic conduits, usually of hard p.v.c., are widely available as an alternative to steel conduits. They cannot, of course, be used as circuit protective conductors as can steel conduits, but their lightness, cheapness and speedy fixing more than repay the need for an additional protective conductor. They have not the same mechanical strength as steel conduits, become soft above 60°C and brittle below −15°C, but are increasingly used for both buried and surface work.

Joining is sometimes by threading, but more usually by a pushfit taper joint, often used in conjunction with an adhesive.

Nonferrous-metal conduits

Aluminium conduits are available, with sherardised threaded inserts suitable for use with the normal steel screwed conduit fittings. Copper and zinc-base-alloy conduits are occasionally used in special situations.

13.11 DUCTS AND TRUNKING

Ducts

A duct is a closed passageway formed within the structure of a building into which cables may be drawn. It may consist of a steel or plastic trunking buried in the floor or ceiling of a building, or may be case *in situ*. This means that a former is placed before concrete is poured, and, when removed after the concrete has set, leaves the duct.

If all or part of a duct is of concrete, only sheathed cables must be drawn into it.

Trunking systems

A trunking is a rectangular-section enclosure, most usually made of sheet steel, but also available in plastic. Where large numbers of cables follow the same route, a trunking system will be used instead of a very large conduit, or a multiple run of smaller conduits, because it is lighter, cheaper, quicker and easier to install. Cables can be laid into the trunking before the lids are closed, instead of being drawn in as with conduits and ducts. Special fittings are available for junctions and bends (Fig. 13.19).

As well as its use for longer runs of cabling, trunking can be installed with advantage to enclose the interconnecting and circuit cables at mains switchgear positions. In some cases, it is necessary to keep circuits separated. This is often necessary where telephone and fire-alarm cables follow the same route as the supply cables; multicomponent trunking can then be used (Fig. 13.20).

13.12 CABLE RATINGS

The rating of a cable is the amount of current it can be allowed to carry continuously without deterioration. The rating depends entirely on the temperature the cable is capable of withstanding without deterioration of its insulation, although, with mineral-insulated cables, the insulation of which will not be impaired no matter how hot it gets, the effect of high temperatures on the seals and on persons touching the sheath is the deciding factor.

Many factors govern the rating of a cable. These are

(a) conductor cross-sectional area: the larger the cable, the more current it will carry without getting hot
(b) insulation: some types of insulation will be damaged at temperatures where no deterioration occurs with others; for instance, a p.v.c. insulated 2·5 mm² cable is rated at 24 A, whereas a mineral-insulated conductor of the same size is rated at 43 A.
(c) ambient temperature: if a cable is installed in a hot situation, it will be unable to dissipate the heat it produces as quickly as it would in cold surroundings, and will operate at a higher temperature.
(d) type of protection: in the event of a fault or of an overload, the speed with which the fuse blows or the circuit breaker opens may differ very widely; if a fault current flows, the temperature of the cable carrying it will increase with time, so the quicker the protective device

186 *Cables and enclosures*

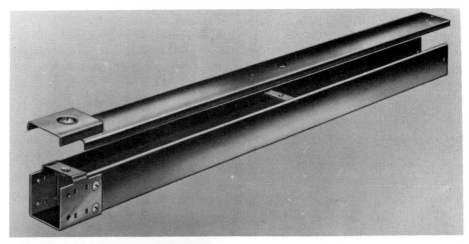

Fig. 13.19 Cable trunking and fittings

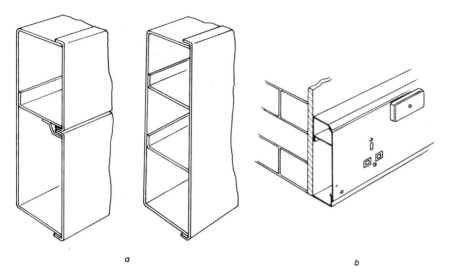

Fig. 13.20 Special trunking
 a Multiple-compartment-type
 b Skirting-type

operates, the less cable heating will occur. Thus, cables protected by rewirable fuses are rated at lower currents than those protected by h.b.c. fuses or circuit breakers

(e) grouping: if a large number of cables is run together, say in a conduit or trunking, they will be unable to radiate, and conduct heat as freely as if they were installed separately. Cables which are in larger groups are hence rated lower than those run singly

(f) disposition: a cable buried direct in moist earth will conduct heat to its surroundings more rapidly than a cable lying in a duct, and can be rated at a higher current

(g) type of sheath: if a cable is armoured with a second overall sheath, it is more effectively heat-insulated than one without armour, and must be given a lower rating if the insulation has the same properties

(h) contact with thermal insulation. A cable in contact with, or buried in, thermal insulation, will be prevented from dissipating heat normally. This situation will often arise in roof spaces, filled cavities of walls, and so on.

The electrical craftsman may have difficulty in deciding what weight to give to each of the above factors. IEE Wiring Regulations provide detailed current-rating tables for all types of cable, together with rating factors to allow for such factors as grouping and disposition. An understanding of these tables is essential for all who must choose a cable for a particular installation.

13.13 EXERCISES

1 List the materials used as conductors for power cables, and compare their advantages.

2 Give the names of *four* materials used to insulate cables, and state where each could be used with advantage.

3 The following materials are used in the manufacture of cables. By giving the important properties of each, show where it could be used.
 (a) copper
 (b) aluminium
 (c) p.v.c.
 (d) p.c.p.
 (e) paper
 (f) glass fibre

4 List *three* situations in which bare conductors may be used.

5 Describe the construction of a twin-with-protective conductor sheathed wiring cable. What are the advantages of this type of cable, and in what type of installation is it likely to be used?

6 Using a sketch, describe the construction of a mineral-insulated cable. Show how this cable is terminated, and explain why a special termination is necessary.

7 List the advantages of mineral-insulated cables. Give two examples of installations where this type of cable would be used.

8 Draw a cross-section to show the construction of a paper-insulated lead-sheathed steel-wire-armoured and -served power cable.

9 Describe the construction of a p.v.c.-insulated and -sheathed steel-tape-armoured cable.

10 Using a drawing, show how a paper-insulated cable is terminated.

11 Describe the compression method of fixing a lug to a conductor, and compare with the sweating method.

12 What precautions must be taken when connecting an aluminium conductor to a brass clamp? Why are these precautions necessary?

13 Use sketches to show how
 (a) two conduits are connected in a coupler
 (b) a conduit is connected in a spout box
 (c) a conduit is connected to a clearance hole.

14 What are the advantages of cable trunking compared with steel conduit?

15 List the factors affecting the rating of a cable. Explain why each factor affects the current the cable may carry continuously.

Chapter 14
Lighting and heating installations

14.1 INTRODUCTION

Having considered the principles of electrical technology, the precautions necessary for safety, and the cables which can be used for installations, we now have to consider the electrical installations themselves. Electrical installations can vary very widely in scope and in complexity, from, for example, a single lighting point on the one hand, to the complete installation of an oil refinery on the other. All installations follow the same basic principles to ensure safety from fire and shock. However, the mains and distribution systems of large industrial systems are complicated, and will not be considered here; domestic and small commerical installations form the majority of the total of electrical installations. Only this type of work is considered in the succeding sections.

14.2 SUPPLY MAINS EQUIPMENT

The supply mains equipment is that normally fixed close to the point where the supply cable enters a building. This equipment is designed to protect the rest of the installation against excessive current, and falls into two categories; that which is the responsibility of the supply authority, and that provided and maintained by the consumer.

Supply authority's mains equipment

The incoming supply cable must be terminated, usually in a sealing box, and the mains cable then feeds to a **main fuse or fuses. The main fuse, or service fuse, almost always of the h.b.c. type, protects the feeder cable** against short circuits at the mains position, and against overloads and short circuits in the consumer's installation. In a properly planned and maintained system, the consumer's fuses or circuit breaker will operate before the main fuse blows; but this fuse is also the insurance by the supply authority against a badly maintained installation. For example, if a consumer's 30 A ring-main fuse blows owing to a fault, and he increases the fuse rating in an attempt to keep the circuit operating, the supply-authority fuse will blow to protect the supply cable.

From the main fuse, the supply feeds on to the meter, which records the energy used so that the consumer can be charged. The sequence of supply equipment is shown in Fig. 14.1.

All of the supply authority's equipment is sealed, because, if connection is made to the supply *before* the meter, the consumer cannot be charged for energy used.

Consumer's supply equipment

The consumer's supply equipment must control and protect his installation against overloads and faults.

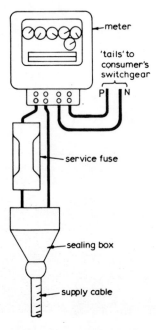

Fig. 14.1 Layout of supply authority's equipment

Equipment is connected to the supply authority's meter in the following sequence:

(a) main switch, which must break both poles of the supply: this means that for the usual single-phase a.c. installation, the switch must be of the double-pole linked type
(b) main fuse: often this main fuse can be omitted, the supply authority's fuse giving protection
(c) distribution board: this board has fuses or circuit breakers to protect the outgoing final circuits.

In most new domestic installations, the supply equipment comprises a consumer's control unit. This unit contains a 60 A double-pole main switch, as well as up to eight fuses or circuit breakers of 45 A, 30 A, 15 A or 5 A rating. The fuse or circuit-breaker rating must not be more than the current rating of the smallest conductor it protects; thus 5 A fuses are used for lighting circuits, 15 A fuses for immersion heaters and similar loads, 30 A fuses for cookers and ring-main socket circuits and 45 A for the larger cookers (Fig. 14.2).

In larger domestic or commercial installations, more than eight circuits will be necessary, and the supply equipment must be larger. There are many possible variations, but a common layout is shown in diagrammatic form in Fig. 14.3.

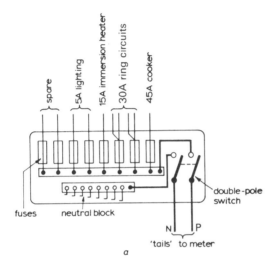

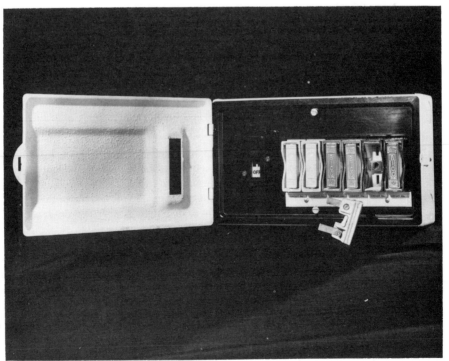

Fig. 14.2 6-way consumers'-service unit

192 Lighting and heating installations

The main switchfuse consists of a linked switch and fuses enclosed in one case. The busbar chamber encloses bare copper or aluminium bars mounted on insulating supports, and is used as a connecting box to couple together the output of the main fuse switch, and the main circuits.

The main circuits are controlled and protected by switchfuses mounted close to the busbar chamber. These main circuits may feed out to individual heavy loads, or may feed distribution fuseboards from which lighting and other final circuits are fed.

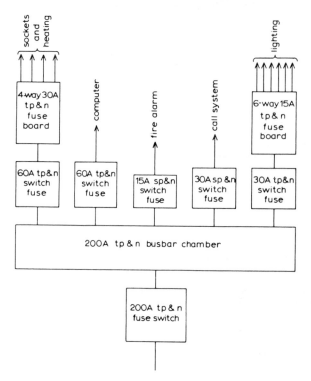

Fig. 14.3 Layout of supply equipment for office block

14.3 LIGHTING CIRCUITS

It is not good practice for a lighting circuit to feed a total load exceeding 15 A. This means up to 36 lights connected to one final subcircuit. In the normal installation, good planning usually limits the number of lights on each circuit to about ten, with more than one lighting circuit to each building. This ensures that the whole of a building is unlikely to be plunged into darkness by the operation of one fuse.

Simple theoretical circuits for the control of lighting points are shown in Fig. 14.4. In practice, the connections used will depend on the wiring system chosen. The loop-in method (Fig. 14.5) is used where cables run through conduit, all connections being made at switches or lighting points. Other joints may be difficult to find in the event of a fault. The conduit can be used to provide the earthing point needed at each outlet if it is of the heavy-gauge-steel type; if light-gauge or plastic, an extra p.v.c.-insulated protective conductor (coloured green) must be used.

Where a sheathed wiring system is used, an included protective conductor will be connected to the earthing points. Single-core sheathed conductors, with the addition of a protective conductor, may be used, connections being the same as for the loop-in system. It is far more common, however, for joint boxes to be used. These boxes may have fixed terminals, or may enclose loose connectors, and are securely fixed to joists under floors or in ceiling spaces. Where situated under floors, they must be covered by a screwed trap in the floor, so that they can be examined if necessary. Connections for the joint-box system are shown in Fig. 14.6.

An alternative system for use with sheathed cables requires a special ceiling rose with an extra terminal, which must be shielded so that it cannot be touched when the cover is removed for replacement of the flexible cord. This precaution is necessary since the third terminal of these '3-plate' ceiling roses is used for

Lighting and heating installations 193

connecting live conductors, and remains live when the controlling switch is off. Connections are shown in Fig. 14.7.

It is sometimes necessary for an existing light controlled by a one-way switch to be modified for 2-way switch control. Fig. 14.8 shows two methods of feeding the extra switch.

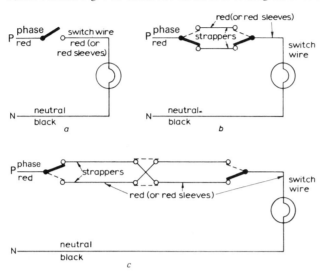

Fig. 14.4 Basic lighting circuits
    a Light controlled by 1-way switch
    b Light controlled by two 2-way switches
    c Light controlled by two 2-way switches and one intermediate switch.
      Light is controlled from any one of three switches
      Alternative switch connections are shown dotted

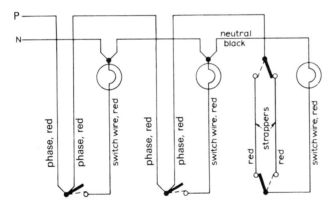

Fig. 14.5 'Loop-in' wiring system. Earth connections at outlets are made to conduit system

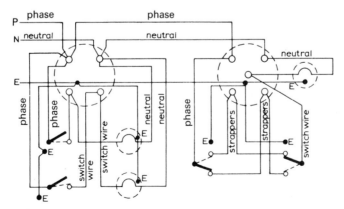

Fig. 14.6 Joint-box wiring system

194   Lighting and heating installations

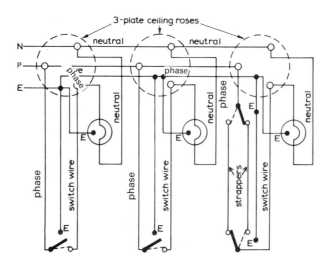

Fig. 14.7   3-plate ceiling-rose wiring system

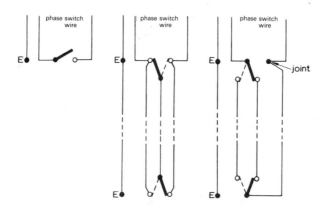

Fig. 14.8   Alternative methods of converting existing 1-way switch to 2-way

14.4   SOCKET-OUTLET CIRCUITS

There are two basic types of socket outlet, and these will be considered separately, since the circuits are quite different.

Sockets for unfused plugs

These sockets accept round-pin plugs which, although available with fuses in the plug, are usually unfused. They are available in 15 A, 5 A and 2 A ratings. Each 15 A socket must be wired back to its own circuit fuse (Fig. 14.9a). 5 A sockets may be looped on a radial final circuit, with not more than three sockets on each circuit (Fig. 14.9b). For three 5 A sockets on one circuit, a 15 A fuse is necessary, so the current rating of the cable used must be 15 A throughout. 2 A sockets are normally used to feed portable lighting fittings, and can conveniently be connected to lighting circuits.

Sockets for fused plugs

These sockets will only accept plugs which are fused. The most common fused plug is the flat-pin type, to BS1363, but other versions, such as the Wylex and the Dorman-Smith, are available.

The fuse in the plug protects the flexible cord and the appliance, so the circuit fuse now protects only the circuit cable and the socket (Fig. 14.10). The system thus has two advantages:

(a)   Since the plug is connected directly to the appliance, the fuse in the plug (13 A or 3 A) can be chosen to give the best protection. For example, a table lamp would be protected by a 3 A fuse, whereas a 2 kW fire would have a 13 A fuse fitted in its plug.

(b)   The number of sockets connected on one circuit can be related to the *total* demand required, without having to assume that each socket always provides its full rated output. Thus, say, six 13 A sockets may be connected to a circuit protected by a 20 A fuse. If each socket were fully loaded, total circuit current would be 6 × 13 A = 78 A, and the circuit fuse would blow, but if all the sockets are in the same room, this is very unlikely to happen. We say that **diversity** can be applied to the circuit.

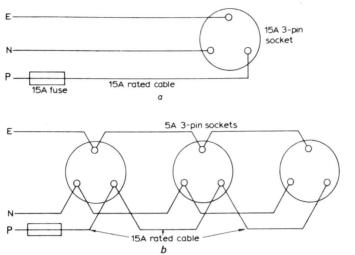

Fig. 14.9   Typical circuits for sockets with unfused plugs

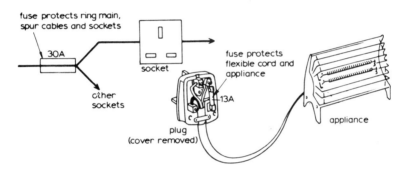

Fig. 14.10   Role of fuses in ring circuit

Fig. 14.11 shows the various circuit arrangements which are permissible with fused-plug sockets, and applied to domestic situations only. A fixed appliance, such as a small water heater, may be connected through a fused spur box, but only one such outlet may be fed from a spur.

The use of fused-plug circuits, especially of the ring circuit (Fig. 14.11c), has permitted the provision of adequate sockets at reasonable cost.

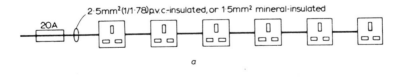

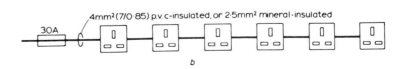

Fig. 14.11   Final circuits for sockets with fused plugs. All connections are 3-core, with phase, neutral and earth connectors.

196 Lighting and heating installations

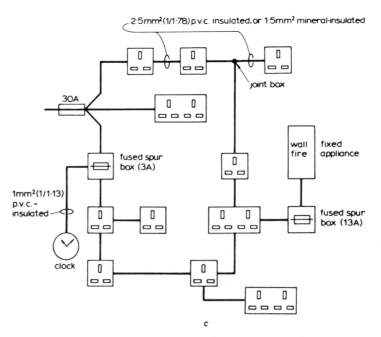

Fig. 14.11 Final circuits for sockets with fused plugs. All connections are 3-core, with phase, neutral and earth conductors.

14.5 OTHER CIRCUITS

The majority of the electrical apparatus in a modern home or office is connected to the lighting circuits, or is fed from sockets. There are, however, equipments where this is either impossible or undesirable. For example, a large electric cooker must be fed back on a separate circuit connected to its own separate fuse, usually rated at 30 A or 45 A. Again, whereas a 3 kW immersion heater may be fed from a ring main, it is often desirable to remove this heavy load from the ring by wiring it back separately to a 15 A fuse.

At the other end of the scale, most houses have an electric bell. The circuit is fed at low voltage from a transformer, usually situated at the mains position. As well as being safe, the wiring can then be carried out using small, easily-hidden and inexpensive bell wire. A typical circuit is shown in Fig. 14.12.

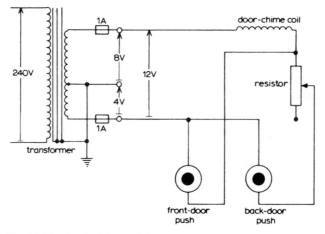

Fig. 14.12 Typical domestic-bell circuit

Alarm Systems
Losses due to fires and break-ins have increased year by year, and the demand for fire and intruder alarms has increased with them.

A simple open circuit alarm system which may be used for fire or intruder purposes is shown in Fig. 14.13. Closing any switch will sound the alarm.

Lighting and heating installations 197

The form of the switches will vary depending on the application. For fire alarms they will consist of 'break glass' points, where a spring-loaded plunger moves out to close contacts when a thin sheet of glass which normally holds it back is broken; or an automatic system will include heat or smoke detectors which are arranged to close on operation. Recent developments in the field of automatic operation in the event of a fire or of a burglary include passive infra-red detectors (which sense the temperature increase due to a fire or to an intruder), ultrasonic and sonic detectors, and microwave detectors. With intruder alarms, it is more difficult to arrange for a switch operated by the unwelcome person to remain closed, and a continuous ringing bell (or a relay to convert a normal bell into one) may be necessary (see Fig. 7.3). A battery, often trickle charged from a mains-fed d.c. supply is preferred to a transformer operating directly from the public supply because of the fear of failure in the event of fire, or of interference by the intruder. The same fears will be felt for the circuitry itself; a cut wire or a burned cable could put the system out of action.

For this reason, **closed-circuit** alarm systems are usually preferred. Instead of contacts which must close to operate the circuit, normally-closed circuit switches are wired in series as shown in Fig. 14.14. The relay contacts which operate the bell are held open by a coil energised through the alarm contacts. A steady small current in this series circuit monitors its condition. In the event of a contact opening, or a deliberate or accidental break in a circuit, the relay will drop and sound the alarm. A pair of auxiliary contacts may be used in conjunction with a telephone line to alert the police or fire service.

This system has much to commend it, because it 'fails to safe'. Fine wire can be stretched across windows and other likely entry points so that they are likely to be broken by illegal entry. Many sophisticated alarms activated by heat, smoke and flames for fire systems, or by body heat, movement, or sound for intruder alarms are not uncommon.

To reduce the risk of fire alarm cables being burned through before they can operate, all such cables must be segregated or separated by fire-proof barriers from all other cables, unless they consist of mineral-insulated cables, which are unaffected by high temperatures.

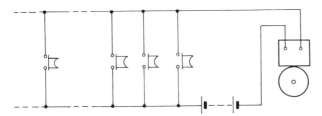

Fig. 14.13 Simple open-circuit alarm system.

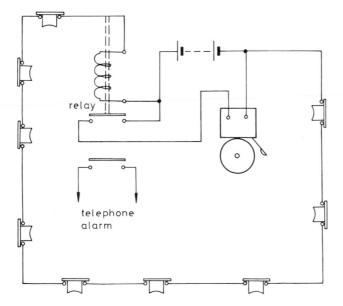

Fig. 14.14 Simple closed-circuit alarm system.

14.6 EARTHING AND POLARITY

The principles of earthing were considered in Chapter 12. Non-current-carrying metalwork must either be protected so that it cannot be touched in the event of a voltage to earth, or connected to earth so that no potential difference can exist between the metalwork and earth. For electrical installations, the method of connecting metalwork with earth has been adopted. The connection is made by joining all such metalwork together by means of the protective conductor, which is itself then connected to earth.

The protective conductor may take a number of forms, such as a separate insulated conductor, the earthing conductor in a sheathed cable, the earthing conductor in a flexible cord (green-yellow core), the sheath and/or armour of a cable, steel conduit or trunking etc. To ensure that lighting switches, lighting fittings and sockets can be effectively earthed, an earthing terminal must be provided in the outlet box serving them (Figs. 14.6 and 14.7). This is necessary even though the switches or fittings are all-insulated; should they at some future time be changed for metal types, earthing could be effectively carried out. The earth connection of a socket must be coupled to the earthing terminal in the socket box by a short earth connector. The connection between the box and the socket by means of the fixing screws is not relied on for earthing purposes.

The importance of polarity has been stressed in Chapter 12. Where single-pole switches are used, care should be taken to ensure that they are connected in the phase (unearthed) conductor (Figs. 14.5, 14.6 and 14.7).

14.7 SIMPLE TESTING

When an installation is complete, it must be inspected to make sure that it complies with the IEE Regulations. This inspection will not show up incorrect or faulty connections, and tests are carried out to eliminate such faults.

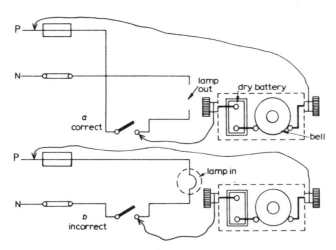

Fig. 14.15, which indicates the method used to test switch polarity
    a Correct test with lamp *out*. Bell rings when one switch terminal is connected to wander lead
    b Incorrect test with lamp *in*. Bell rings although switch is connected in neutral conductor

Continuity tests

The object of a continuity test is to ensure that a conductor is continuous and unbroken along its length. A simple continuity test may be carried out using a bell or buzzer tester, a circuit for which is shown in Fig. 14.15, which indicates the method used to ensure that single-pole switches are connected in the phase conductor.

The simple bell tester may also be used to ensure the continuity of ring circuits, following the method of Fig. 14.16. Three such ring-circuit-continuity tests are necessary, one for the phase conductors, one for the neutral conductors, and one for the earth conductors.

A test is necessary to ensure that the resistance of the protective conductor is sufficiently low to enable an earth fault to blow the fuse or to operate the circuit breaker. Since an actual resistance reading is necessary, the bell tester cannot be used, and an ohmmeter fed from a hand generator or an internal battery must be used. This is connected between a distant point in the installation and the supply position as shown in Fig. 14.17, when it will give a reading of the protective-conductor resistance plus the resistance of the long test lead used. By connecting this lead across the ohmmeter, its resistance can be measured; this value is subtracted from the previous reading to obtain the resistance of the protective conductor.

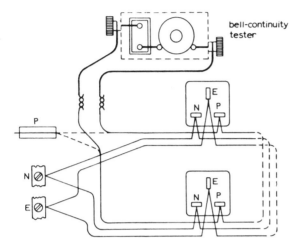

Fig. 14.16 Test of ring-circuit continuity. Phase conductors are shown under test, which will be repeated on neutral and earth conductors

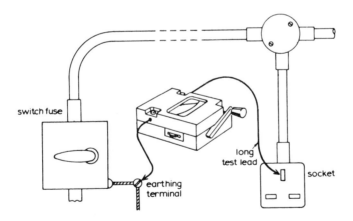

Fig. 14.17 Test of a protective conductor using a hand generator type tester. The conduit shown has a separate protective conductor included; if the separate conductor was not used the conduit resistance would be found using a heavy current tester

Insulation-resistance tests

The insulation resistance of an installation is the resistance from the current-carrying conductors to earth through the insulation. In a sound installation with effective insulation, the value will be high, but dampness, or incorrect connections, or poorly made and insulated joints may reduce it to a low value. The test must be carried out with an instrument having a test p.d. of 500 V, which is obtained from a hand-driven generator or from a battery through a transistorised potential-increasing circuit. Many testers, including those in **Figs. 14.17 and 14.18 are dual-purpose instruments. Operation of a switch converts them from continuity** testers to insulation-resistance testers. A sound installation should have a test value not less than 1 MΩ. A value lower than this indicates ineffective insulation, or may be due to the installation being a large one. In the latter case, more insulation is involved, and since individual insulation resistances are effectively in parallel, the overall resistance may be low. Should this be the case, the large installation should be divided into groups of not less than 50 outlets, each of which should have a measured insulation resistance of not less than 1 MΩ. Fig. 14.18 shows in simple terms how an insulation resistance test is carried out on a lighting installation.

Earth Tests

Tests for earth-fault loop impedance, and earth electrode testing are considered in Section 12.5.

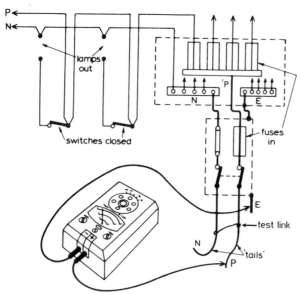

Fig. 14.18 Insulation test from supply 'tails' before connection using battery megohmmeter. Test will be repeated from phase to neutral after removal of the test link

14.8 EXERCISES

1. Write down a list of the items of equipment normally found at the incoming-supply position of an electrical installation, and state the purpose of each.

2. Explain why the main fuse and meter of an installation have seals applied so that interference with them can be detected.

3. Sketch a consumer's service unit, indicating its principal components.

4. Draw circuit diagrams of the wiring connections for
   (a) two lights controlled by one switch
   (b) one light controlled by either of two switches
   (c) two lights controlled by any one of three switches

5. A light in a bedroom is controlled by a one-way switch at the door. Using a wiring diagram, show how the system could be converted for control from the door or from a ceiling switch over the bed.

6. How many lighting points may be connected to one lighting circuit in compliance with IEE Regulations? Would this number of lights normally be wired on one circuit? Give reasons for your answer.

7. Show the connections for two lights, each controlled by one switch, and two lights both controlled by 2-way switches, if they are wired
   (a) using the looping-in method
   (b) in sheathed cable using joint boxes
   (c) in sheathed cable using 3-plate ceiling roses.

8 Draw a circuit diagram to show the mains layout for a socket outlet system comprising 6 – 15 A and 8 – 5 A sockets.

9 What are the advantages of using socket-outlets for which *only* fused plugs are available?

10 Draw a circuit showing the most economical method of wiring 5 – 13 A sockets in a loung of area 24 m^2.

11 (a) How many 13 A sockets may be fed from a ring circuit serving an area of 98 m^2.
   (b) How many spurs are allowed from such a ring circuit?
   (c) How many sockets may be connected to each spur?
   (d) How many fixed appliances may be connected to each spur?

12 What is a protective conductor? What forms may such a conductor take?

13 Why must single-pole switches be connected only in the phase conductor of a system having an earthed neutral? Describe a simple test to ensure that they are so connected.

14 Why must the continuity of a ring circuit be tested, and how can such a test be carried out?

15 Describe how a continuity test of a protective conductor would be carried out, and indicate the results which should be obtained.

16 What is insulation resistance? Describe an insulation resistance test on a lighting circuit, giving the minimum acceptable result.

Chapter 15

Introduction to electronics

15.1 INTRODUCTION

Electronics is that branch of electrical engineering in which valves, transistors and similar devices are used. Not many years ago electronics was a separate subject, studied mainly by those interested in telecommunications, radio, television, etc.

Today, electronics has reached into every field of electrical engineering activity. Motor-control gear, thermostats, boiler-control systems, even the simple door bell, may, and probably does contain electronic components. The electrical craftsman can no longer afford to be ignorant of electronics. This chapter will introduce the subject, and will describe some of the components used.

15.2 RESISTORS FOR ELECTRONIC CIRCUITS

Most of the resistors considered in Chapter 2 are types used in power circuits, which must be accurate and must be capable of carrying heavy currents. In electronic circuits, we are concerned usually with currents of the order of milliamperes, so we are able to accept resistors with low current ratings. Since the currents are low, and because large numbers of resistors are needed, it becomes economical to use small resistors whose value is not so accurately known as is the case with power resistors. For example, a 120 Ω marked 10% resistor may have a value anywhere between 108 Ω and 132 Ω. This variation in values is called the **tolerance**, and is usually given as a percentage (here ±10%).

EXAMPLE 15.1

What are the maximum and minimum acceptable values for a resistor marked at 15 kΩ if its tolerance is
    (a) ±20%, (b) ±10%, (c) ±5%, (d) ±2%, (e) ±1%?

(a) 20% of 15 000 Ω is $\frac{20}{100} \times 15\,000$ ohms = 3000 Ω or 3 kΩ

    Maximum value is thus 15 kΩ + 3 kΩ = 18 kΩ
    Minimum value is thus 15 kΩ − 3 kΩ = 12 kΩ

(b) 10% of 15 000 Ω is $\frac{10}{100} \times 15\,000$ ohms = 1500 Ω or 1·5 kΩ

    Maximum value is thus 15 kΩ + 1·5 kΩ = 16·5 kΩ
    Minimum value is thus 15 kΩ − 1·5 kΩ = 13·5 kΩ

(c) 5% of 15 000 Ω is $\frac{5}{100} \times 15\,000$ ohms = 750 Ω or 0·75 kΩ

    Maximum value is thus 15 kΩ + 0·75 kΩ = 15·75 kΩ
    Minimum value is thus 15 kΩ − 0·75 kΩ = 14·25 kΩ

(d) 2% of 15 000 Ω is $\frac{2}{100} \times 15\,000$ ohms = 300 Ω or 0·3 kΩ

    Maximum value is thus 15 kΩ + 0·3 kΩ = 15·3 kΩ
    Minimum value is thus 15 kΩ − 0·3 kΩ = 14·7 kΩ

(e) 1% of 15 000 Ω is $\frac{1}{100} \times 15\,000$ ohms = 150 Ω or 0·15 kΩ

    Maximum value is thus 15 kΩ + 0·15 kΩ = 15·15 kΩ
    Minimum value is thus 15 kΩ − 0·15 kΩ = 14·85 kΩ.

A very important property of a resistor is its **power rating**. This is the power which may be dissipated in the resistor continuously without it becoming overheated, and depends on the current carried.

$$P = I^2 R, \text{ so } I^2 = \frac{P}{R}$$

Introduction to electronics   205

and
$$I = \sqrt{\frac{P}{R}}$$

where
I = maximum sustained permissible current, A
P = power rating of resistor, W
R = resistance of resistor, Ω.

EXAMPLE 15.2

Calculate the maximum permissible current in a 1 kΩ resistor if it is rated at (a) 0.5 W (b) 1 W (c) 2 W.

(a) $$I = \sqrt{\frac{P}{R}} = \sqrt{\frac{0.5}{1000}} = \sqrt{0.0005} = 0.0224 \text{ A}$$

or 22.4 mA

(b) $$I = \sqrt{\frac{P}{R}} = \sqrt{\frac{1}{1000}} = \sqrt{0.001} = 0.0316 \text{ A}$$

or 31.6 mA

(c) $$I = \sqrt{\frac{P}{R}} = \sqrt{\frac{2}{1000}} = \sqrt{0.002} = 0.0447 \text{ A}$$

or 44.7 mA

EXAMPLE 5.3

What rating should be chosen for a 12 kΩ resistor which is to carry a current of 9 mA?

$$P = I^2 R = \left(\frac{9}{10^3}\right)^2 \times 12 \times 10^3 \text{ watts}$$

$$= \frac{81}{10^6} \times 12 \times 10^3 \text{ watts}$$

$$= \frac{972}{10^3} \text{ watts}$$

$$= 0.972 \text{ W}$$

In practice, the nearest rating above the calculated value would be chosen, i.e. 1 W.

The resistor is often too small for its value and power rating to be printed on it, so a code is used to show its resistance. Two codes are commonly in use:

(a) *Colour code*
A series of bands of colour are printed on the resistor, each colour representing a number, or in some cases a tolerance. The colours, with the values represented, are

black   0
brown   1
red   2
orange 3
yellow 4
green   5
blue   6
violet 7
grey   8
white   9

The colours are applied in four bands (Fig. 15.1). The first band indicates the first figure of the value, the second band the second figure, and the third band the number of noughts to be added. The third band is sometimes coloured gold or silver, indicating one-tenth or one-hundredth, respectively, of the first two digits. Thus when bands appear in the sequence blue-grey-red (in positions 1, 2 and 3, respectively, in

Fig. 15.1), the resistor is 6800 Ω or 6·8 kΩ. Similarly, orange-white-green indicates 3 900 000 Ω, or 3·9 MΩ, and red-red-gold 2·2 Ω.

The fourth band indicates the tolerance, the colour code being

gold ± 5% } not to be confused with gold (× 1/10) and silver (× 1/100)
silver ±10% } when used as the third band.

No fourth band or salmon-pink band ±20%.

Fig. 15.1   Colour coding of resistors

If the bands 1, 2, 3 and 4 are, respectively, red-red-brown-silver, it indicates 220 Ω ± 10%. Sometimes a resistor is specially made to ensure that its resistance does not change as it ages. These 'high-stability' resistors are indicated by a salmon-pink fifth band.

(b) Digital code

This code takes the form of numbers and letters printed on the resistor, and is most often applied to wire-wound types.

R indicates a decimal point, so that 1R0 means 1 Ω, 4R7 means 4·7 Ω, 68R means 68 Ω, 220R means 220 Ω, and so on.

K has a similar function, but indicates values in thousands of ohms, or kilohms, so that 1K0 means 1 kΩ, 4K7 means 4·7 kΩ, 82K means 82 kΩ, and so on.

M serves a similar purpose, but indicates values in millions of ohms, or megohms.

Thus 1M2 means 1·2 MΩ, 15M means 15 MΩ, and so on.

With this code, tolerances are indicated by a code letter placed after the value.

The tolerance code is

B ±0·1%
C ±0·25%
D ±0·5%
F ±1%
G ±2%
J ±5%
K ±10%
M ±20%
N ±30%

Examples of the use of this code are

4R7J   = 4·7 Ω ± 5%
6K8F   = 6·8 kΩ ± 1%
68KK   = 68 kΩ ± 10%
4M7M   = 4·7 MΩ ± 20%

It is intended that this code will eventually replace the colour code.

Many types of resistor are used in electronic circuits:

Carbon-composition resistors are most usually used, and are moulded from carbon compound into a cylindrical shape, connection being by wire ends. They are made in various sizes with power ratings from ⅛ W to 2 W. (Fig. 15.2).

Carbon-film resistors often have higher stability than moulded carbon types. The resistive film or coating is deposited on a glass tube, which is buried in a plastic moulding. The connecting wires are carried into the ends of the glass tube to conduct heat away from the resistor.

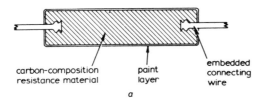

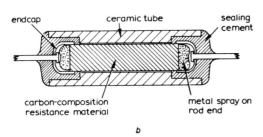

Fig. 15.2 Carbon-composition resistors
  a Uninsulated-type
  b Insulated-type

Cracked-carbon film, or pyrolytic resistors have a film of cracked carbon deposited on a ceramic rod. The film is then cut through in a spiral pattern, producing what is effectively a long thin resistor element wound round the rod. Endcaps with connecting leads, and a coating of silicon lacquer, complete the construction. Cracked-carbon resistors have higher stability than other types.

Metal-film and metal-oxide resistors have a similar construction to the cracked-carbon film type, but the film, and hence the spiral track, are formed of nickel-chromium or a metal oxide. These resistors are capable of operation at very high temperatures.

Wire-wound resistors are used when the power to be dissipated is high. The resistance wire (nickel-chromium) is usually wound on a ceramic tube and given a vitreous enamel coating for insulation and protection (Fig. 15.3).

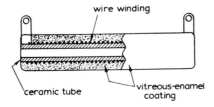

Fig. 15.3 Wirewound vitreous-enamelled resistor

Variable resistors are usually operated by turning a shaft, and are of two types. The wire-wound type is used for lower resistances (up to about $100\,k\Omega$) where higher powers are dissipated. The carbon-track type is used for higher resistances (up to about $2\,M\Omega$).

15.3 SEMICONDUCTOR DIODE

A crystal of the basic semiconductor material, germanium or silicon, has opposite sides treated so that the two halves have different characteristics. One half of the crystal becomes a *p*-type material, with a shortage of mobile electrons, and the other half becomes an *n*-type material, which has a surplus of mobile electrons. Some of these surplus electrons cross the boundary from the *n*-type half to the *p*-type half of the crystal, making the *p*-type material negatively charged, and the *n*-type material positively charged (Fig. 15.4).

Electrons can move easily from the *n*-type region to the *p*-type, but not in the reverse direction. The device thus behaves as a rectifier. The circuit symbol for the semiconductor diode is shown in Fig. 15.5, the arrow of the symbol pointing the direction in which **conventional** current can flow.

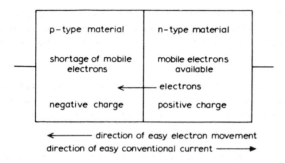

Fig. 15.4 Simple representation of *p-n* junction (semiconductor diode)

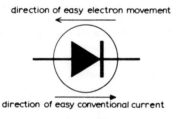

Fig. 15.5 Circuit symbol for semiconductor diode

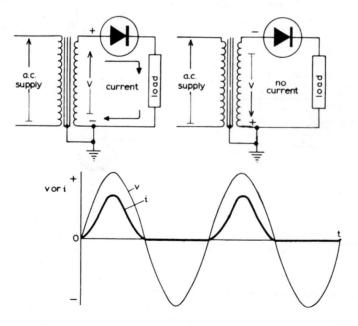

Fig. 15.6 Circuit and wave diagrams for semiconductor diode connected to load as halfwave rectifier

Fig. 15.6 shows a typical rectifying circuit, and indicates how current can pass through the load during the positive halfcycles of an alternating supply, but not during the negative halfcycles.

Semiconductor diodes are used widely in industry, their main application being the conversion of an a.c. supply to give a d.c. supply. Although the action of a single diode does give a direct current, this consists of a series of isolated current pulses, and is quite unlike the output from a d.c. generator or a battery. By using two or more diodes in special circuits, the output may be improved, and the addition of capacitors and chokes allows a smooth supply to be obtained.

15.4 SEMICONDUCTOR-DIODE TYPES

There are four types of semiconductor diode, each having its particular advantages and disadvantages.

Copper-oxide diodes have a low voltage drop when carrying current, but are only able to pass small currents. They are used in conjunction with moving-coil instruments on a.c. supplies. The diodes are arranged in series because a single unit will break down and pass current in the reverse direction if more than about 10 V is applied.

Selenium diodes have a reverse breakdown voltage two or three times that of the copper-oxide unit, but are still commonly connected in series. Current-carrying capacity is quite small, and selenium rectifiers are used mainly for providing comparatively low outputs for such purposes as battery charging and to feed contactor coils.

Germanium diodes have high current rating for comparatively small size, but are becoming less widely used, as failure occurs suddenly if overload raises the junction temperature above about 80°C.

Silicon diodes have very high current rating indeed, and an extremely high reverse breakdown voltage so that series connection is seldom necessary. A silicon diode no larger than a pea can carry a current of 15 A, and one 50 mm in diameter and 40 mm long can handle over 600 A. Fig. 15.7 shows a typical construction for a silicon diode.

These rectifiers must be kept cool to prevent a breakdown, and are often mounted on aluminium castings called **heatsinks** to increase current rating. Very heavily loaded silicon diodes are sometimes water-cooled.

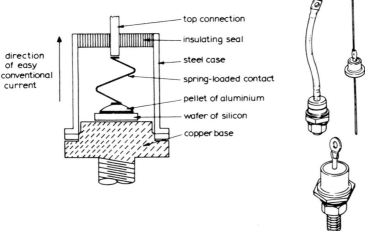

Fig. 15.7 Silicon-diode construction

15.5 EXERCISES

1 What are the possible maximum and minimum values of a resistor marked 680 Ω if its tolerance is
   (a) ±20%,
   (b) ±10%,
   (c) ±5%,
   (d) ±2%?

2   A resistor is marked as 180 kΩ but measured at 192 kΩ. What tolerance does this indicate?

3   Calculate the maximum currents for the following resistors:
- (a) 2·7 kΩ, 0·5 W;
- (b) 68 Ω, 0·125 W;
- (c) 120 kΩ, 1 W;
- (d) 1·2 MΩ, 0·5 W;
- (e) 43 kΩ, 2 W.

4   Give the values and tolerances of the resistors having the following codes:
- (a) orange, orange, yellow, gold
- (b) brown, black, orange, silver
- (c) grey, red, green
- (d) red, violet, gold, silver
- (e) 390RK
- (f) 2K2K
- (g) 3M9M
- (h) 39RG
- (i) 47KJ
- (j) 5K6N

5   Use sketches to assist descriptions of the following types of resistor:
- (a) carbon-composition,
- (b) carbon-film,
- (c) pyrolytic,
- (d) metal-oxide,
- (e) wire-wound.

6   Draw a circuit diagram to show a semiconductor diode connected in series with a load to an a.c. supply. Draw a wave diagram of the supply voltage and the circuit current, describing why the current wave has the form shown.

Numerical answers to exercises

Chapter 1 (Section 1.14)

1. 1200 C
2. 100 min
3. 12·5 A
4. (a) positive, (b) 3·2 A
7. 900 J
8. 25 mm
9. 150 N
10. 44 J
11. 120 000 J
12. 28.9 μV
13. 0·5 A
14. 60 A
15. 4 mA
16. 20 Ω
17. 10 V
18. 15 A
19. 3 A, 15 V, 45 V, 60 V, 120 V.
20. 2 Ω
21. 479·8 Ω
22. (a) 1·5 Ω, (b) 3·6 MΩ, (c) 80 μΩ
23. (a) overload, heating; (b) coulomb; (f) negative, positive; (g) heating; (h) negative; (i) 6 Ω; (j) 5 Ω; (k) current × time; (l) 360 mA; (m) 3 300 V
24. 30 V, 3 A, 3 Ω
25. 0·067 Ω, 6 000 A, 4 000 A, 1 333 A, 667 A
26. 12 V
27. (a) see Fig. 1.11; (b) one half, one quarter
28. 10 A, 6 A, 4 A, 20 A, 2·4 Ω
29. (a) 78 V, (b) 312 V
30. (b) 4·8 A, (c) 8·4 Ω
31. (a) 80 Ω, (b) $I_1 = I_2 = I_3 = 1$ A, $I_4 = 2.25$ A, $I_5 = 0.75$ A, $I_6 = 1.8$ A, $I_7 = 0.9$ A, $I_8 = 0.3$ A (c) $V_1 = V_2 = V_3 = 60$ V, $V_4 = V_5 = 90$ V, $V_6 = V_7 = V_8 = 90$ V
32. 6 Ω
33. 6·7 Ω
34. Bank 1: 30 V, 0·75 A, 1·25 A; Bank 2: 10 V, 2 A; Bank 3: 8 V, 0·5 A in each 16 Ω resistor
35. 10 A

Chapter 2 (Section 2.9)

1. 2·16 Ω
2. 2·5 Ω
3. 0·194 Ω
4. 0·1187 Ω
5. 2·15 Ω
6. 250 m
7. 2 mm²
8. 0·75 Ω
9. 2·5 times
10. 100 Ω
14. 0·342 Ω
15. 3·82 Ω
16. 1·72 Ω
17. 116 m
18. 2500 μΩmm
19. 13·3 mm²
20. 439 m
21. ρ = 16·8 μΩm, so the cable is probably of copper
22. 6·57 mm²
23. 109·9 Ω
24. 1·64 Ω
25. 15·06 Ω
26. 188°C
27. 200°C
28. 3·7 Ω
29. 12·8 Ω
30. 17·0 Ω
31. 9·8 V
32. 2·15 A
33. 3·56 Ω, 4·41 Ω
34. 0·1%
35. (a) 195 V (b) 224 V (c) 234 V (d) 244 V
36. 3 V
37. 236 V
38. 0·25 Ω
39. 4·31 A
40. 234 V
41. 26·9 mm²
42. 35 m
43. 195·9 V

Chapter 3 (Section 3.7)

1. 19 620 N
2. 51·0 kg
3. 200 N
4. (a) 60 000 N/m² (b) 800 N/m²
5. 66·7 N
6. 20 Nm
7. 4200 J, 35 W
8. 22 500 J
9. 4 kW
10. 480 N, 2·4 N/mm²
11. 600 N, 9
12. 375 N
13. 160 N
14. 0·139 m, 222 Nm
15. 500 mm, 500 Nm
16. 130 teeth, 446 rev/min
17. 2800 N, 82°
18. 127° from 6000 N, 10 000 N
19. 600 N
20. (b) 1156 N (c) 1250 N

Chapter 4 (Section 4.8)

1. (a) 333 K (b) 198 K (c) 1273 K
2. (a) 47°C (b) 1227°C (c) −33°C
3. 2·68 MJ
4. 3·76 MJ, 1·04 kWh
5. 9·01 kWh
6. 64·1 MJ
7. 16·7 MJ
8. 8370 J/s
9. 1100 kWh
10. 34·9°C
11. 120°C
12. 14·5°C
13. 27 min 55 s
14. 10·5 kW
15. 369°C

Chapter 5 (Section 5.5)

1. 1·5 kW
2. 48 W
3. 500 W
4. 1200 W
5. 20 A
6. 240 V kettle rated at 1·92 kW, and 200 V kettle at 2 kW
7. 2·4 kW
8. 0·6 A
9. 40 Ω
10. 10 kΩ
11. 3·125 A
12. 10 A
13. 300 V
14. 7·07 mA, 70·7 V
15. 300 J
16. 1·33 kW
17. 200 kWh or 720 MJ
18. 15 000 J

Numerical answers to exercises

19	12 V	26	5·5 Ω		(b) 4·32 kW	34	54 min 16 s
20	£4·80p	27	$P_7 = 175$ W,	30	6·33 A, 29·2 A, 10 A,	35	2·62 kW
21	744 W		$P_5 = 45$ W,		30 A	36	16·4 A
22	2 kW		$P_{10} = 22·5$ W	31	38	37	80%
23	2·7 kW		$P_{30} = 7·5$ W	32	£9·46p	38	21·6 kW
24	20 kW, 83·3 A	28	20 Ω	33	1 kW	39	6 kW, 11·8 kW
25	4·5 kW, £10·80p	29	(a) 12 A, 6 A				

Chapter 6 (Section 6.8)

4	1·3 T	6	(a) 4800 At/m		(c) 6·03 mT	8	965 μWb, 1·21 T
5	6·72 μWb		(b) 4·82 μWb	7	2120 At	9	0·462 mWb

Chapter 8 (Section 8.9)

4	1·4 V	11	3 A, 5·4 V	17	0·4 Ω	22	(a) (i) 1·12 A
5	2 A, 2·1 V	13	27 V	18	1·05 V		(ii) 8·06 V
6	0·2 Ω	14	(a) 2·3 V (b) 3·1 V	19	(a) 90 V		(b) (i) 0·193 A
7	0·0178 Ω	15	(b) (i) 5·5 V, 2 A;		(b) 110 V		(ii) 1·39 V
8	0·025 Ω		(ii) 1·1 V, 10 A	20	1·4 A, 1·26 V		(c) (i) 0·549 A
10	12 V	16	36 V	21	0·4 Ω, 8 V		(ii) 3·95 V
						23	80%, 66·7%

Chapter 9 (Section 9.8)

1	(a) into paper,		(d) right to left	4	2 m/s	7	4 V
	(b) out of paper,	2	2 m	5	0·48 V	8	0·167 s
	(c) north pole at bottom,	3	3 m/s	6	4 V, 800 V	9	60 mWb

Chapter 10 (Section 10.11)

1	1000 Hz	7	127 V, 141 V, 1·11	16	100 V, 60°		(d) 3.18 MΩ, 7·54 μA
2	16·7 ms	8	9·01 A, 14·1 A	17	(a) 12 A, (b) 5·75 A,		(e) 3·98 Ω, 0·251 A
3	(a) 33·3 Hz,	9	216 V, 339 V		(c) 50 A, (d) 10 A,	21	100 turns
	(b) 150 V,	10	191 A, 212 A		(e) 15 A	22	125 V
	(c) 162 V,	11	(b) 70·7 A	18	(a) 314·2 Ω, 0·765 A;	23	1920 turns
	(d) 1·08	12	141 A		(b) 7·54 Ω, 55·4 A;	26	(a) 480 turns
4	141 V, 127 V	13	(b) 7·07 V,		(c) 377 Ω, 0·265 A		(b) 24 V
5	20 ms		(c) 2 cycles	19	(a) 318 Ω, 0·754 A	28	4·17 A
6	(a) 25 Hz, (b) 35 A,	14	(a) 141 A, (b) 135°		(b) 7·96 kΩ, 1·51 mA	29	100 A
	(c) 38·7 A, (d) 1·105,	15	24·3 A leading by 17°		(c) 17·7 Ω, 0·17 A;	30	2500 turns, 1·5 kV
	(e) −40 A						

Chapter 11 (Section 11.8)

1	4·44 A	5	(a) right to left,		(d) north pole at bottom	7	0·25 mA
2	13·3 m		(b) right to left,			8	1·11 T
3	1·1 T		(c) out of the paper,	6	4000 N	9	20 m
						10	12 N

Chapter 14 (Section 14.8)

6	36, no		(b) not more than the number of sockets on the ring itself,		(c) two,
11	(a) unlimited,				(d) one

Chapter 15 (Section 15.5)

1. (a) 816 Ω, 544 Ω;
   (b) 748 Ω, 612 Ω;
   (c) 714 Ω, 646 Ω;
   (d) 693·6 Ω, 666·4 Ω
2. +6·7%
3. (a) 13·6 mA,
   (b) 43·3 mA,
   (c) 2·9 mA,
   (d) 0·65 mA,
   (e) 6·8 mA
4. (a) 330 kΩ ± 5%,
   (b) 10 kΩ ± 10%,
   (c) 8·2 MΩ ± 20%,
   (d) 2·7 Ω ± 5%,
   (e) 390 Ω ± 10%,
   (f) 2·2 kΩ ± 10%,
   (g) 3·9 MΩ ± 20%,
   (h) 39 Ω ± 2%,
   (i) 47 kΩ ± 5%,
   (j) 5·6 kΩ ± 30%

Index

absolute permeability, **85**
air heaters, **72**
alarm system, **196**
alkaline cells, **103**
alternating current, **120, 126**
alternator, **120**
ammeter, **10**
ampere, **4**
ampere-turn, **82**
ampere-hour, **104**
annealed, **174**
arc, **161, 163**
armoured cables, **179, 182**
artificial respiration, **168**
atom, **2**
attraction-type instrument, **97**
average value, **126**

back e.m.f., **122**
battery, **109**
bell, **90**
bimetal strip, **63, 161**
busbar chamber, **192**
busbar trunking, **177**
butyl rubber, **176**
buzzer, **90**

cable rating, **185, 187**
capacitance, **135**
capacitive reactance, **135**
capacitor, **135**
capacity (battery), **104, 111**
cell, alkaline, **103**
cell, dry, **101**
cell, lead-acid, **102**
cell, Léclanché, **100**
cell, mercury, **101**
cell, primary, **100**
cell, secondary, **100**
cell, simple, **100**
Celsius scale, **58**
charge, **3**
chemical effect, **5**
circuit, **3, 10**
circuit breaker, **161**
closed-circuit alarm, **197**
colour code, resistor, **205**
commutator, **121, 149**
compression joint, **181**
conduction, heat, **62**
conductor, **5**
conduit, **183**
consumer's service unit, **191**
contact, direct, **164**
contact, indirect, **165**
contactor, **93**
continuity test, **198**
continuous-ringing bell, **90**
convection, heat, **62**
convector heater, **62, 73**
conventional current, **3**
copper-oxide rectifier, **209**
coulomb, **3**
c.s.p. insulation, **176**

current, **3**
cycle, **126**
cycle per second, **126**

damping, **151**
d.c.-generator principle, **120**
d.c.-motor principle, **149**
delta connection, **156**
density, **43**
depolariser, **101**
dielectric, **135**
digital resistor code, **206**
diode, semiconductor, **208**
direct current, **164**
discrimination, **161**
distribution fuseboard, **192**
diversity, **195**
double-pole switching, **165**
double-pole fusing, **166**
dry cell, **101**
duct, **185**
dynamic induction, **116**

earth concentric system, **176**
earth conductor, **159**
earth electrode, **159**
earth fault loop, **158**
earth leakage circuit breaker, **163**
earthing, **137, 156, 197**
eddy-current loss, **140**
effective value, **127**
efficiency, **45, 111**
electrolyte, **100**
electromagnet, **80**
electromagnetic induction, **116**
electromotive force, **8**
electron, **2**
energy, **7, 44, 66**
equilibrant, **52**
equilibrium, **52**

fahrenheit scale, **58**
fan heater, **73**
farad, **135**
filament lamp, **72**
fire risk, **164**
fire protection, **164**
Fleming's left-hand rule, **147**
Fleming's right-hand rule, **118**
flux density, magnetic, **79**
force, **7, 42**
force on a conductor, **146**
form factor, **129**
frequency, **126**
fuse, **159**
fuse element, **159**
fuse, h.b.c., **160**
fuse, rewirable, **160**
fuse, semi-enclosed, **160**
fusing factor, **161**

germanium diode, **209**
glass-fibre insulation, **176**

hard-drawn, 175
h.b.c. fuse, 160
heat, 58
heat balance, 61
heat sink, 209
heating effect, 5, 72, 158, 163
heating cables, 175
heavy-gauge conduit, 183
henry, 122, 134
hertz, 126
high-stability resistor, 207
h.o.f.r. insulation, 176
Holger-Nielsen method, 168
h.b.c. fuse, 161
hydrometer, 105
hygroscopic, 176, 181
hysteresis loss, 140

indirect contact, 165
induced e.m.f., 122
induction, 116, 122
inductive a.c. circuit, 134
inductive reactance, 134
in situ, 185
instantaneous value, 126
instrument, moving coil, 150
instrument, moving iron, 97
insulation test, 200
insulation, electrical, 6
insulation, heat, 73
insulator, 5
internal resistance, 106

jack, 48
joint box, 178
joule, 7, 44, 58, 66

kelvin, 58
kilo-, xvi
'kiss of life', 169
kumunal, 175

laminations, 140
lead sheathing, 177, 179
lead-acid cell, 102
leakage current, 102
Léclanché cell, 100
Lenz's law, 148
lever, 46
light-gauge conduit, 183
line current, 155
line voltage, 155
local action, 100
looping-in method, 193
losses, 45
loudspeaker, 96

machine, lifting, 46
magnesium oxide, 176
magnetic circuit, 83, 121
magnetic effect, 5
magnetic field, 78
magnetic tripping, 161
magnetising force, 83

magnetomotive force, 82
main fuse, 190
main switch, 166
mass, 42
maximum value, 126
mean value, 126
mechanical advantage, 46
meg-, xvi
megohm meter, 200
mercury cell, 101
micro-, xvi
microphone, 94
milli-, xvi
mineral insulation, 176, 178, 181
molecule, 2
motor, 146
mouth-to-mouth method, 169
moving-coil instrument, 150
moving-iron instrument, 97

nano-, xvi
negative charge, 2
neutral conductor, 154
newton, 7, 42
N-S rule, 82
N-type semiconductor, 208
nucleus, 2

ohm, 9
Ohm's law, 9
open-circuit alarm, 197
overload, 163
overload protection, 160

paper insulation, 176
parallel circuit, 13
parallelogram of forces, 51
p.c.p. insulation, 176
peak value, 127
pendulum indicator, 92
periodic time, 126
permanent magnetism, 78, 86
permeability of free space, 83
permeability, absolute, 85
permeability, relative, 85
phase angle, 131
phase current, 156
phase difference, 131
phase voltage, 156
phasor, 131
pico-, xvi
plastic conduit, 185
polarity, 165
polarisation, 100
polarised bell, 90
positive charge, 2
positive ion, 3
potential difference, 8
power, 7, 45, 60, 66
power rating, 203
pressure, 42
primary cell, 100
primary winding, 137
protective conductor, 158

proton, **2**
P-type semi-conductor, **208**
pulley block, **49**
p.v.c. insulation, **175**

radian, **130**
radiant heater, **62**
radiation, heat, **62**
rating, cable, **187**
reactance, capacitive, **136**
reactance, inductive, **134**
receiver, **95**
reciprocal, **13**
relay, **92**
relative permeability, **85**
rectifier, **210**
regulations, **167**
repulsion-type instrument, **97**
residual current circuit breaker, **163**
resistance, **9, 24**
resistivity, **29**
resistor, **10, 24, 204**
resultant, **52**
rewirable fuse, **160**
rising main, **176**
rocking stretcher, **168**
root-mean-square value, **127**

scalar quantity, **51**
screw rule, **81**
secondary rule, **100**
secondary winding, **137**
selenium rectifier, **209**
self inductance, **122, 134**
semiconductor, **208**
semiconductor diode, **208**
series circuit, **11**
series-parallel circuit, **16**
shock, electric, **165, 169**
short circuit, **165**
silicon diode, **209**
silicon rubber, **176**
single-phase supply, **154**
single-pole system, **165**
single-stroke bell, **90**
simple cell, **100**
sine wave, **129**

socket-outlet circuits, **194**
solenoid, **81**
specific gravity, **105**
specific heat, **59**
specific resistance, **29**
space factor, **183**
square-law scale, **97**
star connection, **155**
static induction, **122**
storage heater, **73**

telephone, **94**
temperature, **58**
temperature coefficient of resistance, **32**
termination, **180**
tesla, **79**
thermal tripping, **161**
thermistor, **34**
thermometer, **33, 58**
thermostat, **63**
three-heat switch, **15**
three-phase supply, **154**
tolerance, **204**
toroid, **83**
torque, **43**
transformer, **137**
trembler bell, **91**
triangle of forces, **51**
trunking, cable, **183, 185, 187**

variable resistor, **27**
vector, **51**
velocity ratio, **47**
volt, **8**
volt drop, **35**
voltmeter, **10**
vulcanised-rubber insulation, **175**

wandering lead, **198**
water heater, **73**
watt, **8, 45, 66, 71**
waveform, **126**
weber, **79**
weber per square meter, **79**
weight, **42**
work, **7, 44**